Design your own
PC Voice Control System
using Microsoft SAPI, Perl and VB

Richard and Darren Harwood

Design your own
PC Voice Control System

using Microsoft SAPI, Perl and VB

Richard and Darren Harwood

Elektor International Media
www.elektor.com

All rights reserved. No part of this book may be reproduced in any material form, including photocopying, or storing in any medium by electronic means and whether or not transiently or incidentally to some other use of this publication, without the written permission of the copyright holder except in accordance with the provisions of the Copyright, Designs and Patents Act 1988 or under the terms of a licence issued by the Copyright Licensing Agency Ltd, 90 Tottenham Court Road, London, England W1P 9HE. Applications for the copyright holder's written permission to reproduce any part of this publication should be addressed to the publishers.

The publishers have used their best efforts in ensuring the correctness of the information contained in this book. They do not assume, and hereby disclaim, any liability to any party for any loss or damage caused by errors or omissions in this book, whether such errors or omissions result from negligence, accident or any other cause.

British Library Cataloguing in Publication Data
A catalogue record for this book is available from the British Library

ISBN 978-1-907920-07-3

Prepress production: Jack Jamar | Graphic Design, Maastricht
First published in the United Kingdom 2011
Printed in the Netherlands by Wilco, Amersfoort
© Elektor International Media BV 2011

119018-1/UK

Table of Contents

Foreword	9
Preface	10
Conventions	11
Trademarks	11
Chapter 1	13
Who should read this book?	14
What will be covered in the book?	14
Further information	14
Chapter 2	15
A brief History of control of computers by voice commands	15
How does speech recognition work?	17
Chapter 3	19
Mains Electricity a healthy respect required	19
Chapter 4	23
What do we want to achieve?	23
So what can we achieve?	24
What skills do I need?	25
What do I need to start?	26
Software Requirements	27
This project uses:-	27
VB6	28
Microsoft SAPI 5.1 SDK	28
Perl(Practical Extraction and Report Language)	29
NotePad ++	31
K8055 Dll	32
Mas.dll	33
Commercial software	33
Audio considerations	33

Table of Contents

Chapter 5 — 35
Grammar and Training — 35

Chapter 6 — 39
Recognition of voice to useful output — 39

Chapter 7 — 41
Perl, a quick primer — 41
Variables and Data Types — 42
Operators — 43
For each loops — 47
foreach — 47

Chapter 8 — 49
Visual Basic Primer — 49

Chapter 9 — 61
The Configuration File — 61

Chapter 10 — 65
The Rachel Program — 65
VB6 Code adding the functionality — 88

Chapter 11 — 105
The Perl Server — 105
So what does the Perl server program actually do? — 140

Chapter 12 — 143
The K8055 (VM110) — 143
The K8056 — 146

Chapter 13 — 149
PC-Control Master and Relay Slave — 149

Chapter 14 — 155
Perl Scripts for switching and talking — 155

Chapter 15 — 157
TTS Voices — 157

Chapter 16 — 159
Making the system human — 159

Chapter 17 — 163
Using our system as a companion — 163
Time — 163
Weather — 165
Shares — 171
How to play music — 173

Chapter 18 — 175
Readily available hardware — 175
Interfacing to hardware K8055 — 175
Interfacing to PC-control Master — 175
Wired or wireless? — 176

Chapter 19 — 177
How to control appliances — 177
How to control gates — 177
How to control doors — 178
How to control lights — 179

Chapter 20 — 181
Can Rachel tell me the temperature and humidity? — 181

Chapter 21 — 185
Can I have my cake and eat it? — 185
Using mobile devices — 185

Chapter 22 — 187
Where can I get Perl? — 187
Where can I get VB6? — 187
Where can I get tts voices? — 187
Where can I get hardware? — 188
Where can I get Winamp? — 189

Index — 191

Table of Contents

Appendix A
199

Appendix B
Perl server listing
201

Appendix C
A sample of our grammar xml file
209

Appendix D
Config File listing
213

Appendix E
Status.dat
215

Appendix F
Stub File
216

Foreword

About the Authors

Richard and Darren Harwood are a father and son team born and bred in darkest Essex, but we will not hold that against them.
Richard has had three careers, the Fire and Rescue Service, followed by several years of Helicopter Piloting, but more importantly 45 years of Electronics, Programming, Invention and Design. *Don't bother doing the math's Some careers were concurrent!*

Darren is a successful computer programming consultant, providing consultancy to many top companies over the years, and currently specializing in scripting and specialist bespoke form handling applications.
Darren has been into computers since he was a child, when his father bought him a computer and gave it with a stern warning," Don't spend too much time on that", so 30 years later!!

Richard and Darren have worked on many projects together, with numerous Patents applied for over the years, aided and abetted by their long suffering wives and partners.
Both are lateral thinkers and each would say that it is their best attribute work wise and has been a real asset on their journey through the University of Life.

And who is Rachel? You will read lots about Rachel in this book. Hmmm, let me describe Rachel … She is quite good looking, she pays attention when spoken to, most of the time, is willing to do most anything (although not always able to comply), never argues, sometimes gives the impression of being a little hard of hearing, talks only when spoken to, generally, never wants to dine out, and best of all comes with an off switch !!*Oops someone has just handed me some divorce papers!*

Ok enough. The system, being mainly computer based, needed to be as human as possible and after a lot of research we found a very, very good synthesized voice, which happened to be called Rachel so we named the system Rachel and it has stuck ever since, but hey, have any voice you want, and call the system anything, Bob, Bill, whatever floats your boat!!

Preface

> Scotty: *"Computer. Computer? Hello, Computer?!"*
> Dr Nicholls: *"Just use the Keyboard."*
> Scotty: *"How quaint!"*
> *Quote from Star trek IV: The Voyage Home*

Computers have come a long way since the sixties but how we communicate with them is still very much in its infancy. We, as humans have been happy to sit and tap keyboards for forty odd years, but as all the old sci-fi movies show, we really want to talk to them and get them to obey our every command !. *Ha fat chance.*
But the technology exists and we can make our computers understand our spoken word, if we throw enough money at the problem we can have a very well behaved, voice controlled telephone system, hell we speak to them on just about every help desk and corporate telephone system!.

But that's not what we want!

We want to control the real things that affect our daily lives, lights, doors, gates, curtains, televisions, music, domestic appliances, the information highway.

Wouldn't that be great?

User: **"Computer turn on hall lights"**   Computer: **"Hall lights on"**

Well read on, by the end of this book you will have the knowledge and detailed instruction on how you can control all of the above at very, very little cost.

Where possible we are going to use technology that is available now in just about every home.

This book will show you how to tie it all together and get the system working, using detailed instructions, diagrams, and code listings.

Are you ready?   **"Engage"**

Conventions

Body text is in Times Roman 10pt, `Program listings are in Courier New 8pt`, *and Authors comments are italicized.*

Tips are in boxes

Double asterisk ** means there is an explanation in the Appendix.

Trademarks

All respective trademarks are acknowledged.

Chapter 1

Who should read this book?

It has to be said from the outset that this is a book for the technically minded, you have got to like computers and electronics to be interested, and if you are not interested you would find it heavy going.

This book is intended for anyone who has an enquiring mind and, at minimum, a basic understanding of electrical and or electronic theory. We are not talking nuclear physicist here, just a basic understanding.

This book is not targeted at any particular age group, but as mains voltages are involved, this project should not be undertaken by persons under the age of eighteen unless supervised at all times when mains voltages are present.

And by the way if you are at Nuclear Physicist level you won't get bored, there are so many variations, tweaks and code changes to keep the most fertile of minds busy and active.
Note: *Voice control of nuclear installations, let me think!, no not a good idea!*

Book Goals

By the end of this book you will be capable of building your own voice controlled system to control: lights, heating, environmental controls, gates, doors, curtains, kettle, music, information highway, security, health monitoring etc, etc. this list and possibilities are endless.

Not only will you be able to control many and various pieces of equipment but your system will talk back to you in one of the most human of synthesized voices available on the market.

The ultimate goal is to provide, if you will, a digital companion who will respond to your spoken word and tell you it has done so.

Best of all it is not an expensive exercise, we have deliberately selected software and hardware that is cheap but does not compromise on the end result of 97-99% speech recognition.

You could spend thousands on getting the results we can show you for the cost of this book , makes

Chapter 1

this book seem a bargain; must speak to my agent!!

Has that wetted your appetite?

What will be covered in the book?

The book will cover every aspect you need to know, from understanding how speech recognition is achieved to hardware selection, software selection, interface selection to interface design and system build.

You will be guided through setup of hardware and software and there will be primers to get you started on the various programming languages involved, and no, you don't need to be a programmer, we have supplied all the functional code needed but we will teach you enough to enable you to alter the software to meet your needs.

We've supplied URL links to various sites, where you can download software, utilities, suppliers etc.

The Elektor support website for this book

Its waiting for you at Elektor.com. Simply search for this book. There you will also find a support website with details of:
 – Coding and software
 – URL's(web page shortcuts)** to suppliers
 – An interactive FAQ
 – Knowledgebase

Further information

Go to www.harwoodrandd.com

Chapter 2

A brief History of control of computers by voice commands

So where did it start?

Well it had to start when telephony came of age; back in the 1870's when Alexander Graham Bell discovered how to convert air pressure waves to electrical waves. This was the fundamental step in getting the spoken word to be able communicate with electronics of any type. At this stage there was no voice recognition on the horizon, but in the 1950's Bell laboratories developed a working speech recognizer that could recognize numbers only. We are talking 80 years after Bell's discovery.

It is interesting to note that much of the funding for Automatic speech recognition was funded by the US military and in the 80's by the US defence Department. It is amazing how funds being introduced by a government body bring a sinister connotation to the concept!

The first practical device was the IBM "Shoebox" device which was shown to the Public at the 1962 world fair, but that too could recognize only numbers. So named because it was about the same size as a shoe box (not bad size wise, for the era), now this device was not a computer as we know it today, it was in fact a series of audio filters, each recognizing a pitch, high, medium and low.

It worked by recognizing the fact that numbers follow a frequency pattern for example "zero" has a "high, middle low" signature whilst say a "five" had a "high, middle, high" signature.
The output from these filters was fed through rudimentary hardware decoders. So that, for any one signature pattern it lit a single light bulb, not exactly any practical purpose but none the less a demonstration of what was to come.

We must bear in mind that computers were pivotal to the effective recognition of speech and they had been developing in parallel since early 1939 with the founding of HP and their audio oscillator device. Followed in 1940 with Bell Telephone Laboratories CNC (Complex Number Calculator), then in 1943 Konrad Zuse finished his Z3 computer which was, a mix of electronic and mechanical components with over 2000 relays.

The first computer as we know it today was the ENIAC in 1946, which could perform 5000 operations per second, but on the downside it required 1000 square feet of floor space and several opera-

Chapter 2

tors to keep it alive, and even then it only managed 39 hours a week of uptime. About the same as an average human worker today!

But real work on voice recognitions didn't start until the birth of the personal computer.

The PC as we now know it started off with the likes of IBM whose (much changed) PC standard we all pretty much use today, early pioneers like the Spectrum ZX80 (1980 by Science of Cam-bridge Ltd. (later to be better known as Sinclair Research), Commodore(1982), Compaq (1983), Apple(1984), Amstrad and all the other makes that followed.

Back to the speech recognition history...

In the early 1970's The ARPA (Advanced Research Projects Agency (US)) speech understanding project was formed, with the goal of making speech recognition the understanding of speech not merely the recognition of speech. This project led to much cross fertilization of ideas and projects and got speech recognition "Out there".

During the 1980's there were a number of products being developed which went down one of two avenues. The first avenue was the development of speech recognition for telephone systems; the other was for the recognition of the spoken word, for the main purpose of dictation.
These early systems could not recognize continuous speech, they in fact recognized one word at a time, and the speaker had to pause in between each word to get a simulated command to work.

Now, the advent of the PC in the 80's, fuelled by better operating systems got the PC to the stage where it was possible for anyone to program. That's when the flood gates opened and we saw many, many applications being developed for speech recognition, mainly in the dictation and telephony automation fields.

Voice recognition for the control of electronics had been seen by the bigger manufactures to be a waste of their time, as they could not see a large potential market. *Big Mistake, Huge!!*
However, in 1982, Ray Kurzweil started an American company called Kurzweil Applied Intelligence Inc. with the goal of producing a voice activated word processor. Indeed in 1985 his company produced a system that could effectively recognize 1000 words, this was a landmark event, and they went on to produce a 20000 word recognizer in 1987

In July 1997 Lernout and Hauspie, a Belgium based company, acquired Kurzweil and developed the systems further. That same year Microsoft came to the party and invested over $40 million US with their company to investigate the SAPI (Speech Application Programming Interface) compliant speech synthesis and recognition. That technology is now integrated into Microsoft operating systems. Indeed this technology is the one we use in this study.

Great so we now have a system that works….Well nothing is 100% right, is it?
So even today, throughout the entire speech recognition industry, the goal of 97% accuracy still remains. The last elusive 3% is going to be a hard battle indeed, mainly recognizing the many Chinese dialects.

So there we are; up to date,

We end up with hundreds of speech recognition systems on various operating systems and languages,

So let's get started and see which suits us and what these can do for us.

How does speech recognition work?

Your voice is fed into the computer via a microphone and sound card; the sound waves are converted within the soundcard to a digital representation of the sound waves. These wave files are then available to the PC's operating system where the files are then analyzed.
The first pass looks for Phonemes.

What the hell is a Phoneme?

Phonemes are linguistic units. They are sounds that when grouped together form words.
How a phoneme converts into sound depends on factors such as the surrounding phonemes, speaker's sex, age, continental and regional accent.

Examples of Phonemes are:

A	Sound AAH	Cat			
E	Sound EEH	peg	Bread		
I	Sound IE	pig	Wanted		
O	Sound OH	log	Want		
U	Sound UH	plug	Love		
ae	Sound AI	pain	day	gate	station

So they are, if you will, the building blocks of words.

The computer when analyzing the word 'wave file' looks for all the phonemes usually using one or both of the following techniques.

Fourier Transform** which looks at frequency against amplitude, and then compares the pattern against a template of a phoneme, or by using a Markov Model.**

Think of a Markov Model (in a speech recognition context) as a chain of phonemes that represent a word. The chain can branch, and if it does, it is balanced.

Take a look at this diagram of a Markov model of the word Tomato broken down into phonemes.

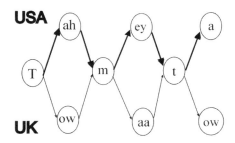

You will see that different phonemes produce the same word. However, one is how the English would pronounce it and the other is how the Americans would pronounce it; either way it's still a tomato, right!

So having got our phonemes, it's pretty simple for the PC to do a second pass consisting of a look up of a vocabulary of words that contain those phonemes to find the closest match. *Houston we have recognition!* Job done …word found!

> NOTE: We refer to those vocabularies as Grammar in our program.
>
> We list only words we want to recognize and place them in the context tree we want them to be activated under.

Chapter 3

Mains Electricity a healthy respect required

Now you don't often get a book that starts out by giving you a warning about the commodity you are going to be working with, but this project and many like it use mains voltages, these voltages where ever in the world they be are lethal if you come into physical contact with them.
So what makes mains voltages lethal?
Well you must ideally understand how mains power is transmitted from power stations to our houses. A much simplified explanation is: - Power generated at power stations is generated in 3 Phase , that is, three power supplies are generated simultaneously albeit each out of phase with the other. This requires six cables coming from the power station right? No wrong! Early experimenter and \inventers found that the earth is a pretty good conductor of electricity. Water, minerals and salts make the earth a good conductor of electricity.
Enter the accountants to the equation…. Why have six cables coming from the power station when you could have three?
Only three!
How does that work?
Well each phase has a 'live' cable to the generating unit and a return' cable (it is technically arbitrary since the electricity produced is an alternating current and it reverses direction every 50th of a second in Europe and every 60th second USA). For infrastructure protection, a large copper plate is buried at the power station and it also provides a return connection to earth. Just prior to this, the power has gone through a transformer to elevate the voltage to several hundreds of thousands of volts for transmission along the national grid pylons (the higher voltage makes transmission losses much lower and allows for thinner cables). At the consumer end of the grid is another transformer or series of transformers to reduce the voltage to our domestic level, which in Europe is anywhere between 210 and 240 volts and in US 105 to 120 volts.

Chapter 3

At this low voltage transformer there is one (two or three) live cable connections. The return cable is also connected at the transformer to a metal plate buried in the earth to form a fault current return path.

We as consumers see generally three cables coming into our homes, which we know as Live , Neutral and Earth (some area's have a fourth which is used for protective multiple earthing where the ground is particularly non-conductive).

So electricity flows from live to neutral? Yes and no, we have already said that it's an alternating current so it is constantly flowing in both directions; however, we can assume that the live conductor is the one carrying the potential voltage, and that it requires a return path.

Well here's the rub - the return path can be earth, and I don't mean the earth cable I mean anything at earth potential, the ground, your floor, indeed just about anything you touch!

This is where the danger lies.

If you make physical contact with the live wire and you are dressed normally with feet on the ground or any part of your body touching any object connected to earth then you're likely going to get a shock!

The severity of the shock is governed by a multitude of factors which include but are not limited to, the insulation properties of the shoes you are wearing, the type and thickness of the carpet you're standing on, whether you are sweating or not, and the amount of skin in contact with the live component.

Make no mistake all shocks are bad for you, from a tingle to a jolt that throws you across the room.

Shocks cause various effects on the human body:

- **Burns**… The heating caused by the body's natural resistance causes deep, extensive and difficult to treat burns. Generally the higher the voltage the worse the burn, due to greater current flow. Typically voltages above 400 cause the worst burns, and yes, you can find voltages at that level in the home and workplace, 415 volts and more exist in three phase installations.
- **Ventricular fibrillation**…this is where your heart starts beating irregularly resulting in heart failure. Mains voltages from 40 volts upwards across the chest for a fraction of a second will cause ventricular fibrillation.
- **Neurological effects**… Current can cause interference with your nervous system, particularly in the head and even lungs, causing the classic stunning effect. So all in all, we need to avoid getting a shock at all costs. It is worth noting that DC shocks are considerably worse than their equivalent voltage AC shocks. This is due to the fact that the potential is fluctuating between 0 and 240 volts AC whereas it is constant at 240volts DC. Dc also tends to cause muscle spasm that can cause you to grip the offending conductor involuntarily and not be able to release it, thus prolonging the period of electricity flow through the body.

OK some simple rules to bear in mind.

- Don't touch any live circuit or any circuit you think to be switched on. Even if you think it is off, if the mains plug is not out then assume it's not safe.
- Never use two hands on live electrical circuits, keep one hand in your pocket, you at least will not get a shock directly across the chest.
- Whilst working on a circuit with one hand, don't use the other hand to hold a chassis, metal cabinet or lean against a wall or other earth path. In fact no exposed skin should be allowed in contact with any item at earth potential.
- Never assume an item is turned off, unplug it yourself.
- If the mains switch is located remotely place a notice, clearly marking that switch "**Do Not Turn On, deliberately turned off, due to persons working on the circuit**", or words to that effect.
- Never assume components in a circuit are not live.
- Never work on electrical item with bare or damp feet.
- Never allow water or liquids near live circuits.
- Always protect circuits with good insulated cases.
- Never use water based fire extinguishers on electrical fires.
- Never overload a circuit beyond its capacity.
- If you are unsure about a circuit, don't work on it, get a qualified technician.
- Never, wear metallic jewellery when working on electrical or electronic circuits.
- Never, use multiple extension cords as a permanent feature (more of a fire risk than that of shock).

Chapter 3

- Always use insulated tools and treat everything on a circuit board as if it were live.
- If you are under 18 years old don't touch any exposed electronics without them being disconnected from mains and a parent being present and fully supervising.

Now, one last warning!

Just because your electronic device is battery powered do not think that it's safe.

Even the humble laptop generates voltages in the thousands of volts at amperages that can kill (e.g. power supply for flat screen backlighting). If it has a battery, disconnect it or remove it first before making adjustments near live conductors or components.

The above is neither exhaustive or a complete or authorative reference, it is provided to remind you that if you connect to mains voltages you must know what you are doing and make sure you're not distracted by any external event, including *"dinners ready" !!!*

Chapter 4

What do we want to achieve?

Ok let's start off in an ideal world, *"Computer iron my shirt"*; not going to happen any time soon!! "Computer turn on the lights", = achievable, with 99% success rate.

"Darling wife would you please iron my shirt so that I can come shopping with you ", has been known to work, but again limited to about **97**% success rate and without the "*so that I can I come shopping with you*", **0**% success rate!

So yes, we want to be able to give any command, request, whatever, vocally. Reality is that we don't yet have the automation available to carry out some of the more human physical actions, or, if it is available, it's too busy making cars and certainly not affordable just yet.
Furthermore, the speech recognition available to us at little cost is limited to about 85-97% consistent success (more if you spend the time training the system).

But setting our sights just a tad lower yields some interesting possibilities. We want to control our immediate environment; now that covers a great deal, for example, we may want to control our home environment in the following areas:

- Heating
- Lighting
- Watering (as in garden)
- Access (as in doors and gates)
- Health monitoring
- Multi-media playing
- Certain domestic appliances
- Communications
- Information Highway (World Wide Web) access
- Spoken reports on weather, shares, news, emails, in fact just about anything that is streamed or available to listen to or read from the internet.

Here is a schematic showing connection via K8055/K8056 boards and by PC-Control master slave setup. We have detailed the K8055 setup but have included the PC-control version, for comparison or where you really want to control equipment up to 1kilometre away.

Chapter 4

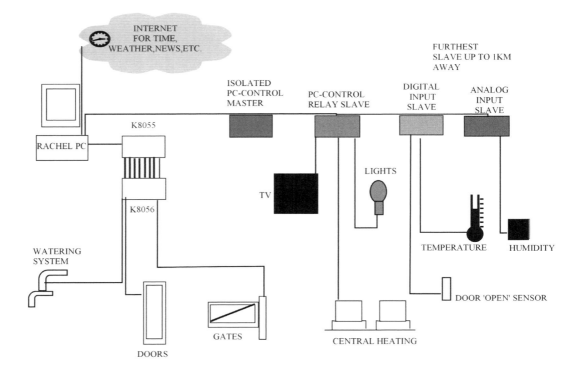

There's a lot more that we can add but, you're going to have to wait for PC Voice Control System vol. 2 for that.

So what can we achieve?

We can achieve everything we want from the environmental list! The capabilities of the system we are going to describe include the following, but remember you don't have to use all or any of these capabilities. Just build, configure, and use the ones you want.
- The system will be capable of fully controlling your heating.
- The system could fully control lighting, "on", "off", "dim", "sequence", "time on", "time off".
- The system could automatically, or under single command, water your garden, greenhouse, flower boxes etc.
- The system could fully control Gates, "open", "shut", "lock", "unlock". Doors, "lock", "unlock",

- Health monitoring, here is a wacky one, you could monitor whether a person responds vocally within a predetermined period and cause an alarm if no response is heard!! The system can provide spoken reminders of when to take medicines (we don't advocate use of this system for any medication taken that life is dependent upon. It is just a thought provoking suggestion).
- Multimedia, the system can fully control media players; loading music by title, artist, album, "volume up", "volume down", "next ", "previous", "pause".
- Domestic appliances? Due to space limitations in this volume we are going to limit domestic appliances to controlling a power socket, "on" or "off".
- Communications? The system could send a sms, email, initiate or end a phone call
- The system can provide spoken details about the current weather, shares, time and date,

Will the system really do all this?

Yes indeed! We personally have this system running in our home office doing all the above tasks and more, and I have to say generally all commands are carried out correctly first time.

Our system has been named "Rachel", and will be henceforth referred to as "Rachel" during the rest of this book.

You have no idea what pleasure it gives us to get up from our desks at close of day and say "Rachel I am leaving the office" and "she" politely says "Goodnight Guys I am turning the lights off for you", which "she" then does , in sequence, between our desk and the door downstairs, as we walk out.

What skills do I need?

I touched upon 'required skill' a little earlier in the book, but the biggest skill required is the desire to learn sufficient about this subject. We can, and will, guide you through the entire project.

If you have programming skills you will find any modification of our software easier than a non-programmer, of course. Similarly, electronics buffs are going to design and build their own interface and not buy prebuilt. But I assure you, you can get this system working without prior knowledge of electronics or computer programming. *Even our crash test dummy has done it (apologies to the wife)!*

You do, however, need to know how to use a PC. In this respect you should be competent to load programs, browse the internet, and generally manage your computer and connections to it.
Ok! I know I am going on about it but again I must remind you about mains voltages and safety.

Chapter 4

Mains voltages can kill! If you are not competent and realistically confident around mains voltages then get a fully qualified electrician to do the mains voltage parts.

If you are under 18 you should have an adult supervising you on this project, especially when it comes to connecting to any mains appliance (We would add that you can do an awful lot of this project and get it to respond to your spoken word without involving mains connections). We further recommend the software part to any teenager, it gives a great sense of achievement, controlling something, *doesn't it girls!?*

Ok lecture over, coffee time.

What do I need to start?

Firstly we need a personal computer or laptop.

Sorry Mac owners we have not catered for you in this volume and the system will not run from a Mac as such unless you are running "Windows Operating System" under "Parallels" or similar VM **software.

The PC's specification is not too critical but there are two factors that really affect speech recognition.
- Number one is that speech recognition is a cpu intensive process, so if we want continuous speech recognition we want a fast PC to do the number crunching as it were. Anything above 2GHz should be fine, fast is good, and faster is better!
- Number two; we need it to be able to respond to the outside world as fast as possible so we are looking for a PC with a minimum of usb2.0 standard for its usb ports.
- By virtue of the above it must have USB ports available. The usual adage, loads of memory is good, more memory is better; it always works to keep things flowing.
- We require a good soundcard; this is pivotal to our input and output. If we have a cheap soundcard we cannot expect it to perform as well as a better quality one. Remember we need good clear spoken instructions that the soundcard can digitize to get through to our speech recognition software.
- Along with the soundcard we need a microphone; again choice is down to the level of accuracy that is required. I use a Bluetooth headset, but a good quality desktop microphone will work fine, one with a ptt (press to talk) ensures that ambient sounds don't get taken as commands, or a good quality headset. Just one further comment on mics, if you're using a laptop the built in mic usually wont hack it; get an external and pop it into your mic-in jack. A CD/DVD / or usb stick is required for getting programs etc onto the PC.

That just about covers the immediate hardware requirements for getting started. As we progress you will have to build or purchase an interface kit or prebuilt unit. They are not expensive and are available as, item number K8055 in kit form from Velleman together with the K8056 relay kit. Both are available as kits or prebuilt. Or you can use the PC_CONTROL Master Slave units, these units are prebuilt only.

If you are using this as a serious system then you would of course additionally consider uninterruptable power supplies and schedule regular backups.

Software Requirements

The speech recognition we are going to describe uses the Windows Operating System. So which one? Well if you want rock solid and easy to use then get a copy or licence for XP Professional.
I know you're going to say we are living in the dark ages, but in this application that is simply not true, if you can't or won't locate a copy of XP then it will have to be Vista or Windows 7. You may have to run programs in compatibility mode or as administrator.

Ok this is where the fun begins.

This project uses:

- Microsoft Visual basic VB6 (you may have a job installing and running that on vista or Windows 7 as they are not yet officially compatible)
- Microsoft SAPI 5.1 sdk**
- ActivePerl 5.2 for our Perl Scripts
- Notepad++
- A mas.dll if we are using PC-control master /slave units
- A K8055.dll for controlling the Velleman K8055 if we are using that interface board.

Nothing too bad about that list and I would expect a lot of you have most of those programs installed already.

Now if you have them all installed then you can jump to the next chapter because I am going to talk you through the install and setup of all the software right now, with the exception of your base operating system.

VB6

Right, let's start with VB6, the first hurdle is getting a copy of VB6 as it has been around for quite some time. You can use VB2010 from Visual Studio, but the going gets tough, as the code we supply may have to undergo conversion. So best bet is go to ebay or similar and do a search under Microsoft VB6, you will usually find a number of copies for sale.

There is no magic to the install, just let it run and install the defaults to default locations, job done, well almost. Now do a reboot of your PC, Microsoft os likes that .

We use VB6 to call SAPI and Visual Basic makes this a real easy task.

Microsoft SAPI 5.1 SDK

SAPI is the Speech engine we are going to be using so the easiest way to get all the bits at one go is to download the Software Development Kit, free from Microsoft

This development kit can be downloaded from http://www.microsoft.com/downloads/en/details.aspx?FamilyID=5e86ec97-40a7-453f-b0ee-6583171b4530&DisplayLang=en

Once downloaded install the SDK to its defaults. We will utilize the sapi.dll to do some of the basic voice recognition. There is nothing for us to set up; our program will call the various components it needs from their default location.

The SDK contains the following which are of interest to us:
- API definition files as C or C++ header files.
- Runtime component the sapi.dll.
- Control Panel application to select and configure default speech recogniser.
- Text-To-Speech engine
- Speech Recognition engine
- Redistributable components to allow developers to package the engines and runtimes with their application code to produce a single installable application.
- Sample application code.
- Sample engines - implementations of the engine interfaces.

Perl (Practical Extraction and Report Language)

Perl is the Perl of programming languages. *That'll start an argument!*

Yes we love Perl and not just because it's free, but also it is a very versatile programming language and it is very easy to learn.

Perl is, technically a dynamic interpreted language originally written by Larry Wall in 1985 to produce reports and text based processing, on Unix computers. The program consists of a core and then it is supplemented by a vast and ever increasing range of add-on modules, which gives it its strength and flexibility. This language is now considered to be in the public domain (Part of the Open Source software movement- distributed under the GNU Public License.).

Many of the add-on modules are written by enthusiasts all over the world and are held on repositories in universities worldwide.

Thousands of websites use Perl for CGI scripting, but its use is not limited to web applications by a long way, as you will see later on in the book.

So where do you get it from?

Perl is available free from Activestate.com for the community edition, you can find it here at this URL http://www.activestate.com/activeperl/downloads

You would normally download the x86 version for Windows, especially as we are supporting only Windows in this book. Once downloaded install it, let it install to its default locations, and although there is no good reason to, then reboot!!

Ok downloaded and installed? Almost done...

Go to Start menu, select Active Perl when you will find folders called Documentation and Perl Package Manager. The Perl Package Manager is where we get and install the vitally important modules that we are going to use, so go ahead and **click** the Perl Package Manager to open it. Now **select** Edit, Preferences

Chapter 4

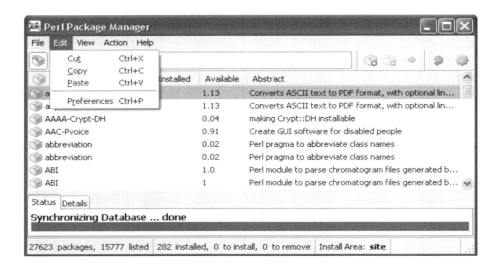

Then click on Repositories, you will see the following

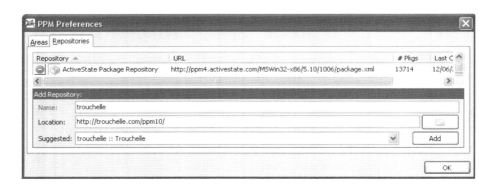

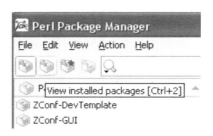

This will list all the modules you have installed, scroll down until you see the items listed under win32 and look for this specific line Win32::OLE. If you don't see it **click** on the next icon to the left, View All Packages, scroll down until you find Win32::OLE, then right **click** on it and select Install Win32::OLE, then **click** Action at the top of the page and select Install Win32 OLE.

There are other packages that you will need to install to make the ser-dch3.pl run and the same goes for the time and weather scripts, but now you know how to do it. To test our installation, type the following into Notepad ++ application - Print "Hello World\n";

and **Save As** hello.pl in the perl install folder of your PC, then go to Start Menu, type in **cmd** in the Run box and press **enter**. Use the DOS cd command to navigate to your Perl installation directory and then **type** perl hello.pl and press **enter**. If all is working well you should see Perl respond with Hello World, in the command console window.

You have successfully installed and set up Perl! Just to make life easy take a look at.

NotePad ++

Notepad ++ is a super text editor ideal for source code editing, it supports many languages as well as our Perl script and xml script writing. Notepad ++ is very versatile and comes with a host of text editing features which makes it ideal for us to use, and hey ho! It's free from http://notepad-plus-plus.org/download

Just download and install, and let it do it's own install, accept any defaults offered.

One of the features I like is that you can have as many notes as you like open at the same time under tabs. You can close the program and when you reopen the program they are still there where you left them, great !!!

It has built in intelligence, which is a great help whilst editing, give it a try.

K8055 Dll

This dll comes with the Velleman kit and is used in their demo software.

The K8055 is a proprietary Dynamic Link Library supplied free from Velleman to support their K8055 USB experimenter's board. You can download it from here http://www.vellemanusa.com/us/enu/download/files/ Scroll down the list until you see the K8055 Dll revision 3 and download it. It is of course supplied with their k8055 kit or ready built unit.

The files come in a zip file which you will have to extract to your hard drive.

You should **save** the dll to your c:\windows\system32 folder. If you do, you then have to register it on your PC, and this is not to register as in send info off to some website, it is to register it as in, meaning make your computer aware of its location on your pc.

To register the dll:

Open a command prompt window, by **selecting** Start, Run, and then type in "cmd" and press **enter**. Using the DOS change directory command navigate to the windows 32 folder where you saved the dll to.

Then **type** into the command box   "regsvr32.ex k8055.dl" and press **enter**. The dll will then be registered and available to your Microsoft and Perl applications that use the system environmental path.

We can then call the various functions we need to control the k8055 unit.

> TIP!
> OK not withstanding all of the above, which is the correct way?
> Quick and dirty, you can just copy the K8055.dll to your Perl directory and make sure it's in the same folder as your perl scripts, and it will function perfectly, your choice!!

Mas.dll

Again like the k8055 dll this mas. dll is supplied with PC-control master boards. This is the dll that supports the PC-control master slave units. This dll doesn't seem to like to be registered so just ensure it is copied to the directory that your perl scripts reside in, and then it will run, no problem.

The dll can be obtained from www.pc-control.co.uk it is as we stated earlier also included with any purchase of their Master control units or from the Elektor support site

Commercial software

We could go out there and buy or acquire any one of several speech recognition programs, such as Nuance Dragon Naturally Speaking, Via Voice (no longer supported for windows versions above Vista), Tazti (free), CMU Sphinx (open source free), but we will not be doing ourselves any favours at this stage of the game. The commercial systems are built generally with a single purpose in mind, mainly dictation with a few computer commands bolted in to allow you to carry out windows functions like Scroll, New, and Start Menu etc.

They do not give us access to the **result** of a word command other than their pre ordained action.

If we are to get speech recognition under our control then we need to access the **result** of a spoken word and be able to translate that into useable code to carry out the functions we need via a PC interface board to achieve our goals.

Audio considerations

Pay attention guys this bit is important. Most, if not all PC's come with some kind of sound card of varying quality, *you pay your money and you take your pick* but you need to know that all soundcards are not born equal.

We don't require any special features but the standard features have to be of good quality, that is to say we want our spoken word to reach the SAPI without distortion or interference. So any soundcard will do the job, but as with many things if we want really good recognition, a better quality card may be in order.

Chapter 4

We are not so interested in whether it does surround sound or fibre out, we are interested in the fact that the components of the card, are well shielded from interference from within the computer case, and that it faithfully reproduces the spoken sound (good signal to noise ratio), and has decent amplifiers built in to provide sufficient gain to any input microphone we provide.

Whilst we are talking about microphones, (*this opens up a whole new can of worms*), if we want super accurate recognition then we do want the best we can afford.

So what are we looking for in a microphone? Well, faithful reproduction is number one and we are not going to get it from a low end microphone, so go for a mid-priced microphone. Ideally you want noise cancelling, unidirectional (that is, it has to be pointed at the speaker (as in person, not item attached to PC), why? because we don't want background noise that could degrade our recognition performance). It can be a headset, or even wireless, but if it's wireless you do lose control over what you can do *pre PC*, and although it's nice to just talk without pressing buttons a press to talk (PTT) really does cut out false recognitions.

Pre PC?, What I mean by that is that if you really want terrific results then you are going to put a microphone preamp in between the microphone and the PC, this allows us to set the perfect audio level, but with wireless or USB microphones we don't (easily and economically) have the option of connecting a preamp.

On a normal sound card you will not have much in the way of a microphone preamp, as it is only designed for low quality speech and cheaper end microphones – not really suited for serious speech recognition.

You can add Microphone Preamps to your set up via any of these methods:
- Mixer (e.g. Behringer UB 802)
- Sound Card with built in Preamp (e.g. Tascam US-122)
- M Mixer (e.g. Behringer UB 802)
- Maplin   Microphone Preamplifier Order Code: QS41U

They will all help in making the spoken word easier to recognize. So the ideal setup is: Good quality microphone connected to PC via a microphone preamp or amplified mixer.

I use a mixer in which I can plug in multiple microphones so that Rachel can be talked to from just about anywhere in the office.

Chapter 5

Grammar and Training

No, we are not going to give you an English lesson, this grammar is the list of words that your speech recognition system will learn and more importantly how it will treat each word by applying "Rules" and a hierarchy to those Rules and words contained within them.

SAPI inherently ships with a grammar built in, but that grammar is directed at controlling a PC, the windows applications and operating system.

We, however, need a number of specific words and rules to be recognized and they are not contained within the standard grammar. Luckily for us, the designers of SAPI recognized that users may want specific grammars loaded and have made provision for this.

Default Grammar is loaded into the SAPI system upon start up but later you have the opportunity to substitute the built in grammar for your own. But beware the grammar MUST follow strict parameters, otherwise it will fail and nothing will be recognised.

The grammar is stored in an XML file (Extensible Markup Language). XML is a form of computer readable scripting if you will that is capable of passing parameters to a host program. It is not unlike HTML in appearance with tags surrounding keywords.

This   <!——means ignore, it's a comment and you terminate it with-->

Further reading: - http://msdn.microsoft.com/en-us/library/ms723635%28v=VS.85%29.aspx

The program simply sits there like a lemon if you don't get the grammar right. But worry not, all you have to do is write what you want in the config file (see Chapter 9.), in plain English. We have written an interpreter contained within the Perl Server that will create the xml for you and put it in the correct place.

Chapter 5

Our Grammar, looks like this:

```
<GRAMMAR LANGID="409">
<DEFINE>
        <ID NAME="DEF1071" VAL="1071" /> <!-- Dummy0 +on -->
        <ID NAME="DEF1070" VAL="1070" /> <!-- Dummy0 +off -->
        <ID NAME="DEF1061" VAL="1061" /> <!-- Dummy1 +on -->
        <ID NAME="DEF1060" VAL="1060" /> <!-- Dummy1 +off -->
        <ID NAME="DEF1051" VAL="1051" /> <!-- Pool Table Area +on -->
        <ID NAME="DEF1050" VAL="1050" /> <!-- Pool Table Area +off -->
        <ID NAME="DEF1041" VAL="1041" /> <!-- Arcade Area +on -->
        <ID NAME="DEF1040" VAL="1040" /> <!-- Arcade Area +off -->
        <ID NAME="DEF1031" VAL="1031" /> <!-- Bottom of stairs +on -->
        <ID NAME="DEF1030" VAL="1030" /> <!-- Bottom of stairs +off -->
   <ID NAME="CmdType"  VAL="0001"/><!-- VAL  digit 0 =ignore-->
   <ID NAME="Commands" VAL="0002"/><!-- VAL first digit 0 =ignore-->

        </DEFINE>
        <RULE ID="Commands" TOPLEVEL="ACTIVE">
        <P>+Rachel</P>
        <RULEREF REFID="CmdType" />
        </RULE>
        <RULE ID="CmdType" >
        <L PROPID="CmdType">
        <P VAL="DEF1071">Dummy0 +on</P>
        <P VAL="DEF1070">Dummy0 +off</P>
        <P VAL="DEF1061">Dummy1 +on</P>
        <P VAL="DEF1060">Dummy1 +off</P>
        <P VAL="DEF1051">Pool Table Area +on</P>
        <P VAL="DEF1050">Pool Table Area +off</P>
        <P VAL="DEF1041">Arcade Area +on</P>
        <P VAL="DEF1040">Arcade Area +off</P>
        <P VAL="DEF1031">Bottom of stairs +on</P>



        </L>
        </RULE>
</GRAMMAR>
```

Grammar and Training

Get the picture? Don't try and copy the grammar above, it is only a snippet so it is incomplete, but it does give you an insight to the complexity of the xml grammar.

This grammar can be written using notepad or notepad++ and you must copy the structure exactly if/ when manually producing the grammar xml.
Ours is a very simple but fully functional Grammar.

It does not have Rules per say, it is more a list of Phrases we want to be recognised and the only true rule is that we make it 'Require" the word Rachel before it will listen to anything else.

This is to enable us to have a normal conversation without the lights going on and off or the gates opening, or Rachel reading us the latest weather forecast!

For example "weather "won't get a response, but "Rachel weather" will get recognised and a Value returned to the VB6 program along with the recognised phrase.

So EVERY command that we want recognised MUST start with the spoken word, "Rachel". *OK Heads up! If your house is full of women with the name Rachel, choose another name for the Keyword else life will get tedious.*

Any word that we feel is very important can be accentuated by placing a + in front of it, the program then gives more importance to that word.

Now you can get very, very sophisticated with these grammar xml's, for example you can set a Rule that has the Keyword (Top Level) music and make it only respond to words that are music specific, such as volume up, next, etc. You can then, if you like, create a mini vocabulary based under a Keyword and only words that are specific to the keyword are acted upon.

To understand this a bit more you might want to divert to Chapter 9 and read about the config file that we use to automatically build the xml grammar file. Be sure and come back to here though!

Chapter 6

Recognition of voice to useful output

Our system gets speech from the microphone via the soundcard, using our VB6 Program.
SAPI "listens" to the soundcard interface to see if it can "hear" a Keyword (Top Level in our XML Grammar File), if it "hears" a Keyword it then listens for Phrases that we have entered into the Grammar file.

In our case the SAPI system listens for the Keyword "Rachel", then it "listens" for a phrase such as "Office light one on". It does this by a kind of sort system, for example;
- Did I hear a Keyword? No, continue listening.
- Did I hear a Keyword? Yes, (Rachel)
- Did I hear any first word of any phrase in my Grammar file? Yes, (Office)
- Does it match an entire phrase in my Grammar file? No, continue
- Did I hear any second word of any phrase in my Grammar file? Yes, (Light)
- Does it match an entire phrase in my Grammar file? No, continue
- Did I hear any third word of any phrase in my Grammar file? Yes, (one)
- Does it match an entire phrase in my Grammar file? No, continue
- Did I hear any fourth word of any phrase in my Grammar file? Yes (on)

Does it match an entire phrase in my Grammar file? Yes, Generate Hypothesis, (if No then go back to listening for Keyword)

Our VB6 Program Indicates Recognition (turns the events Frame Green), gets the numerical "Value" of the recognised Phrase from Grammar file, then sends that value to the Perl Server script, to take action.

The Perl Server script looks at the returned "Value" and then either sends a command to the relevant computer interface board (K8055 unit (or PC-control master unit)) or calls an internet enquiry script, which goes to the Internet, grabs the latest weather or whatever and then reads it aloud.

> Note: On all SAPI driven speech recognition engines there is limited recognition out of the box.
> Only if you train the system will it recognize your particular voice with good reliable accuracy.

Chapter 6

Finally the Perl Server script returns a success "value" to the VB6 program, and invokes an audio script to announce that the required command has been executed.

For example in the scenario above, Rachel would say aloud after she had recognised the Phrase "Office light one has been turned on".

The VB6 program then updates its graphics GUI showing that the command has been completed. Control is then returned to the listening part of the VB program. This loop as it were (it isn't a loop, it is Event driven) continues until the system is turned off.

Chapter 7

Perl, a quick primer

Perl is known as an "Interpreted" language, which is parsed twice. First parse, the syntax is checked and compiled into Byte code, and then it runs. So it is in reality an Interpreted/ Compiled language.

> **TIP!**
> Perl is a scripted language, which means you don't need any special IDE to work with it, in fact most programmers prefer to use Notepad++ to work with Perl.

Now, is Perl a script or a program? Well a program must run through a compiler to produce either assembler or byte code, whereas a script is usually run through an Interpreter. Perl does a bit of both, so none but the purists will be upset if you use either term. (Perl Scripts can easily be changed into program "exe's" , so that they can Run as stand- alone applications).

Perl is a command line program, it runs in a command line window generally, and doesn't generally produce any fancy graphics, so all you're likely to see are results and only then if you have programmed the script to display results.

Whenever you begin to write a Perl Script, you start off with:
#! /usr/bin/perl...Well that's only strictly true on UNIX environments, but you will see it at the top of most perl scripts. The reason you see it on most scripts is that an awful lot of perl scripts are written for internet webpage support, and a lot of those internet hosts use unix servers, or an os based on Unix- such as Linux.

The #! Is known as (hash-bang or shebang) *don't go there! I have no idea why it has those names,* but its function is based on UNIX type operating systems, in which the interpreter parses the rest of the first line as an Interpreter Directive.

Because the "#" character is used as the comment marker in this and many other scripting languages, the contents of the shebang line will be automatically ignored, for interpreter purposes, apart from using the rest of the line as a location path to the Perl executable.

Chapter 7

> **Fact:**
> Most Perl code is platform-independent; it can run on almost any operating system

In the example above it means the Perl executable is located in /usr/bin/.
So that's the shebang line sorted, next!
We just said that the "#" hash is regarded as a comment, so anything you write on a line after"#" will be ignored by the interpreter. If your comment exceeds one line make sure you start off the continuation line with another hash sign. Get used to commenting every line as you go, it really helps when looking at the script later, and you can at least follow the logic that was whizzing around inside your head when you wrote it!

Variables and Data Types

Variables don't require to be Initialised before use. If a variable is referenced before it has been given a value, it will return 0 (zero) or "" an empty string.
Variables are Global by nature.

Variables do not have to be declared before using them, unless you use the "use strict;" pragma (it is a compiler directive) and is placed at the beginning of the script. That then forces all variables to be declared with "my" operator before being used.
It is said to speedup program execution and it definitely helps to find typo's ($steel instead of $stel for example). I recommend that you always use the "use strict" directive.

Variables begin with special characters ($, @, or %), they can consist of letters, underscores and numbers, in any combination, up to 255 in length.

Examples:

```
Scalars begin with $:              $blue_eggs   = 1;
There are no character variables, just use string    $a = "Blue"
Arrays begin with @:                     @ages = (10, 11, 12);
Hashes begin with %.  %employees = (100 => "George", 1033 => "Trevor");
```

There are no "Int" type variables; all numbers are stored as type "float". Variables are typeless; there is no distinction between float and string, that is to say:
- $a= "hello world"; concatenation $x= $a.$a produces "hello worldhello world".
- $b=10; concatenation $c= $a.$a; produces "1010" , $c=$a+$a; produces "2".
- $x= $a+$b; produces "10", this is because a string ($a) is treated as a zero.
- $x=$a produces a compiler error, why? Because you MUST terminate a statement with a semi colon";"

Operators

Operator	Example	result	Definition
+	8+8	= 16	Addition
-	8-2	= 6	Subtraction
*	8*8	= 64	Multiply
/	8/2	= 4	Division
==			Equal to
!=			Not equal to
>			Greater than
<			Less than
>=			Greater than or Equal to
<=			Less than or equal to

One of the best things about Perl is CPAN.
CPAN is the Comprehensive Perl Archive Network.
It is where you go to find Perl module scripts
There's nothing particularly special about Perl modules. They're written in plain text files which you can edit or create in most any text editor. Perl modules typically have a .pm file extension instead of .pl.

Chapter 7

eq	if ($i eq "coal")		Equal to (string)
ne	if ($i ne "coal")		Not equal to (string)
gt	if ($i eq "coal")	Compares alphabetically	Greater than (string)

> **TIP!**
> Perl starts execution at the top of a program (outside a subroutine) and progresses downwards

lt		Compares alphabetically	Less than (string)
ge			Greater than or Equal to (string)
le			Less than or Equal to (string)
&&			AND

||			OR
!			NOT
.	$a="dog"."cat";	= "dogcat"	Concatenation

Output commands

print "hello\n";   #prints hello and sends end of line hidden character.
             #The \n being the end of line instruction

Perl if statement

The perl if statement is structured thus:

If (condition) { Then}

For example
if ($x==10) {
print "hello"
}

If $x does not equal 10 the "if" is ignored.

Using else

```
if (some expression) {
    true statement;
} else {
    false statement;
}
```

Using elsif

```
if (some expression) {
    true statement;
} elsif (another expression) {
    another truestatement;
} else {
    false statement;

}
```

Perl subroutines can be defined as either *named* or *anonymous* subroutines..
A named subroutine can be declared and defined anywhere in a Perl script, using the syntax:

Sub say hello();

The structure is *NAME, BLOCK*

Where NAME is the name you select for your subroutine, and *BLOCK* is the actual programming logic, surrounded by curly braces ({ . . . }).

When selecting names for subroutines, it's recommended that you use all lower case names; since, by convention, subroutine names with all capital letters indicate actions that are triggered automatically as needed by Perl; such as AUTOLOAD or DESTROY
Here's an example of a basic subroutine definition, as well as the code we would use to execute (call) it: ignore the line numbers they are not used in perl, they are there to show the individual lines of perl and nothing else.

Chapter 7

1	`say_hello();`
2	
3	`sub say_hello {`
4	`    print "Hello, World!\n";`
5	`}`

Most times that we call a subroutine we do so because we are trying to reduce code repetition or do a repetitive calculation.
When using subroutines we can pass data in and out of the subroutine itself.

While loops

While loops continually iterate as long as the conditional statement remains true.
Caution: It is dead easy to write a conditional statement that will run forever and never come out, endless especially at the beginner level of coding. Try and give yourself some kind of escape route.

The syntax is while (conditional statement) {execute code; }.

```
# SET SOME VARIABLE
$X = 0;

# RUN The While loop
while ($x <= 5) {
        # PRINT THE VARIABLE
        print "$x\n";
        # INCREMENT THE VARIABLE EACH TIME
        $x ++;
}
print "Finished printing x\n";
```

For loops

A for loop counts through a range of numbers, running a block of code each time it iterates through the loop.

The syntax is:

for($start_num, Range, $increment) { code to execute }.

A for loop needs 3 items placed inside of the conditional statement to be successful.
First a starting point, then a range operator, and finally the incrementing value, for example.

```
for ($i = 10; $i >= 1; $i--) {
        print "$i ";
 }
 print "Finnished.\n";
```

For each loops

foreach

This statement takes a list of values and assigns them one at a time to a scalar variable.

Each time executing a block of code.

```
foreach $i (@some_list) {
        statement1;
        statement2;
    }
$i +=2
```

The following example takes in element and mutliplies it by 3 and replaces the original values in the array. For example;

```
@a = (4, 5, 8,9);
   foreach $one (@a) {
       $one   *= 3;
   }
```

@a now equals = 12, 15, 24, 27

This is a useful function when using arrays.

Chapter 8

Visual Basic Primer

What is Visual Basic? Well it's the most popular and most widely used programming language in the history of computer programming. It started out life as plain old BASIC a non GUI (Graphical User Interface) programming application.)riginally credited as a creation of two Dartmouth (USA) College professors John Kemeny and Thomas Kurtz. Later a company called Tripod developed a forms application, which wasn't associated with programming per say, but produced a pleasant graphical interface. So Microsoft made a deal with Tripod and they bolted the two products together and produced a graphical version of BASIC, codenamed Ruby (not the ruby programming language we know today). That was the birth, in 1991, of Visual Basic 1.0 known universally as VB. It was the first truly graphical programming language that allowed people to easily write programs for the Windows operating system, whilst inherently producing the "Windows" look and feel.

VB6 came with its own IDE (Integrated Desktop Environment), which meant everything you needed to write, build and run your programs was all in one desktop view. It was truly a first with pre made objects that "knew" about windows behaviors. The objects, known as components, could be dragged from a toolbar selector and placed on a form. These components had functionality built in, so that you could program with the minimum of code.

Everything that you program in VB has the feel of windows, and it all integrates seamlessly. So no wonder it was, and still is a favorite amongst programmers, and for beginners to the world of programming there is nothing better.

Many top level programmers scoff at VB because it is simple to use and learn. However, just because it is easy or simple to learn doesn't mean it is not a powerful and practical language.

Coming right up to date, VB is still there in the form of vb.net contained in Microsoft Visual Studio 2010.

VB Net is light years away from standard VB6. It is a far more capable programming language and has a vast array of connection possibilities all designed to get connections with other networks, more web based if you will.

Enough history for one day, let's go programming, and get the feel for VB. If you are already a VB guru, skip this chapter then head on back to get some advanced stuff with the actual Rachel program.

Chapter 8

We will start with the time honoured program "hello world" (fact or fiction nearly all programming languages start by first teaching you how to output the text "hello world" as your first application).

Open up your copy of VB6 and you will be presented with The IDE and a popup window offering you a selection to choose from, as to what kind of program you wish to make. We will need to select the standard exe, as that is what we are about to create.

Your screen will look something like this, depending on the screen resolution you have set. 1280 x 1024 gives a reasonable screen. Anything less will look a tad shabby.

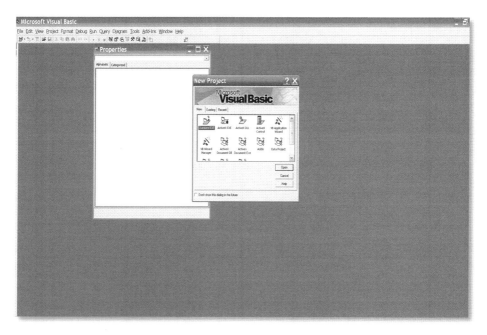

Select Standard exe.

The IDE will then open up a programming interface with a Form (a form is like a blank canvas; if you will, something to place your components on). You will see Project 1 and, in front of that, Form 1. Believe it or not, what you have there is a fully working windows application. If we wanted to stop there it would compile and then display a blank Form on the screen. The form wouldn't do much but it is functional and thus can be Maximized and Minimized and Closed, not bad for zero code eh?

To continue our "hello world" program we are going to select a component from the Component list on left hand side of the IDE.

The component we are looking for is called a Text box. Locate it, click and drag it onto the Form1 form. You will need to resize it to the desired shape and size.

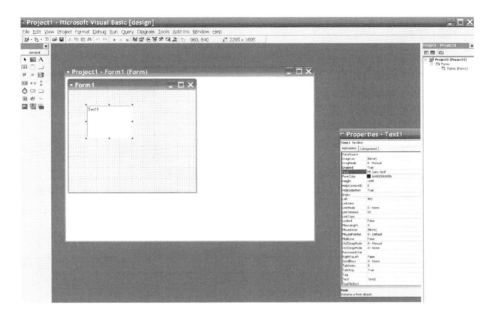

As implied by its name this component can, amongst other things, contain Text. You will notice on the right hand side of the IDE there is a box entitled Properties. These are the user accessible properties that the text box exposes, so that you can customize the component at Run time (when the program is running as opposed to when you are writing the program).

Scroll down the Properties box until you see the line Text, you will see Text1 in the box beside it, highlight that text and replace it with hello world.

You will see in your form that the text inside the text box has changed to hello world.

Chapter 8

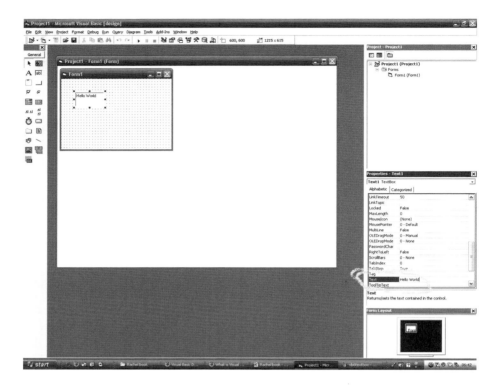

All this coding is so tedious!

I don't think so, you've written exactly two words.

Don't worry you will get plenty of opportunity to "write" code when we get to the Perl side of things; boy you will then long for "Visual" programming languages.

Now click on the little triangle at the top of the IDE. This is the Run button;clicking it both compiles your program and Runs it in Windows.

Visual Basic Primer

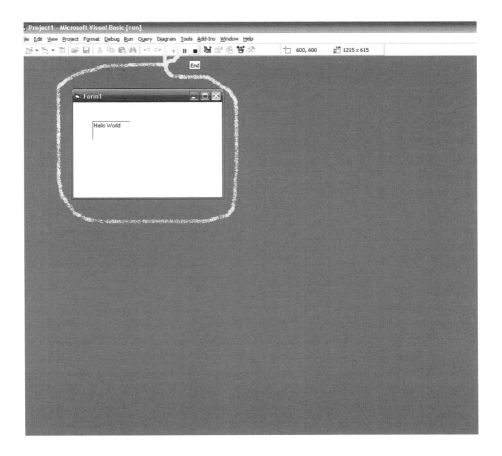

Your new program runs, and is an application in its own right; you have just programmed in VB.

No point in rushing off and showing your partner as they won't be as impressed as you are just yet, but they will be when you're finished.

Now then, what about some code? This visual stuff is ok but where is the code?

Actually The code IS there but its hidden. You can, if you wish, show code and object by having panes, but I find it a distraction. So one or other works for me. To see the code at the top of the IDE is a Menu line, click on View and Select code.

Chapter 8

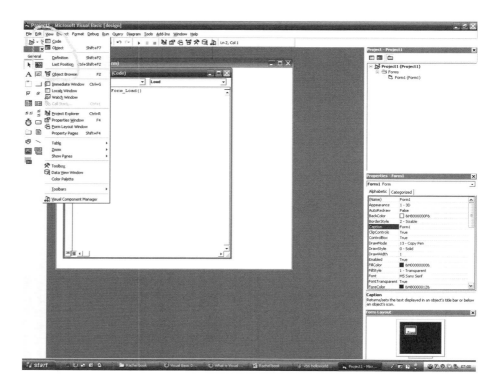

You will see a form open up and all it says is:

"Private Sub Form_Load() End Sub"
That can't be it!
But it is, since Visual Basic is "Visual" by nature.

The components "Know what to do", it's built into them, so you only have to provide interaction and modify the components behaviors.
Now you know why we are using VB.
But wait there's more!!!

To enable VB to be up there and powerful, not only can it use components, but it can use, very powerful libraries of prepackaged code and these libraries can be "Referenced" from within your program, effectively including all the functions contained in the libraries into your program.

We will be using such libraries in the "Rachel Program".

Visual Basic Primer

Also there are other pre-packed "objects" called Active X controls. These are small program building blocks using reusable code. These save you from reinventing the wheel each time you want to use a particular piece of functionality.

Ok, that's a bit about the visual side of VB. What about the code?
Well its very easy; here a few of the normal commands and statements.

Dim statements

```
Dim - Used to define a variable as a certain type

   * i = dim i as integer, r as single
```

You can use the Option Explicit to make sure that VB forces you to declare every variable you use. DIM is the simplest way to declare a variable

```
ReDim - Used to change the dimensions of a dynamic array

   * redim arrayname(12)
```
You can use ReDim to create an array whose size grows by 1 every time you want to add a number to it. Then, use UBound to tell you how many numbers you've added.

```
Static - Establishes a procedure variable which keeps its value between calls

   * static i as integer
```

For example, if you want to keep track of how many times you've been in a procedure set a counter as STATIC and increment it by one for each visit to the procedure. It will never go away until the program is terminated.

```
Public - Creates a variable which can be accessed outside its own procedure

   * public i as integer
```
Even if you're the only programmer writing code in your application, use of Private vs. Public will help catch errors if you inadvertently try to access an out-of-scope variable

Chapter 8

```
Private - Creates a variable that can be read only in its own procedure or module,
according to where the declaration took place.
    * private i as integer
```

Use this as often as possible to avoid unnecessary exposure of your variables to coding mistakes.

Operators

```
/ - Normal division
\ - Integer division (truncates the answer)
^ - Exponentiation operator
* - Multiply
+ - Plus
- - Minus
= - Equal
> - Greater Than
< - Less Than
<> - Not Equal
>= - Greater than or equal
<= - Less than or equal
```

String Functions

```
left$ - Returns the left n characters of a string

* temp$ = left$ ( teststring$, 4 )

right$ - Returns the right n characters of a string

    * temp$ = right$ ( teststring$, 4 )

trim$ - Removes leading and trailing spaces of a string

    * temp$ = trim$ ( teststring$ )
mid$ - Returns n characters from a string, starting a any position

    * temp$ = mid$ ( teststring$, 1, 4 )
```

ltrim$ - Removes only the leading spaces of a string

temp$ - Removes only the trailing spaces of a string

   * temp$ = rtrim$ (teststring$)

len$ - Returns the length of a string (how many characters it has)

   * temp$ = len (teststring$)
ucase$ - Makes all characters upper case

   * temp$ = ucase$ (teststring$)

lcase$ - Makes all characters lower case

   * temp$ = lcase$ (teststring$)

If and buts

```
If..Then..Else -
```
Performs code based on the results of a test
```
        If A>5 Then Print "A is a bit number!"
    For...Next -
```
Loops a specified number of times
```
        For i = 1 to 5: print #1, i: next i
    For Each ... Next -
```
Walks through a collection
```
        For Each X in Form1.controls: Next X
    While..Wend -
```
Loops until an event is false
```
        while i < 5: i = i +1: wend
    Select Case -
```
Takes an action based on a value of a parameter
```
        select case i
        case 1 : print "it was a 1"
        case 2 : print "it was a 2"
        end select
    Do...Loop -
```
Loops until conditions are met
```
        do while i < 5 : i = i + 1 : loop

    Choose -
```
Selects and returns a value from a list of arguments
```
        Choose (index, "answer1", "answer2", "answer3")
    With -
```
Executes a series of statements on a single object
```
        With textbox1
        .Height = 100
```

Chapter 8

```
            .Width = 500
        End With
```
`End` – Immediately stops execution of a program
```
        End
```
`Stop` – Pauses execution of a program (can restart without loss of data)
```
        Stop
```

`GoTo` – Switches execution to a new line in the code
```
        GoTo Line1
```
`GoSub . Return` – Switches execution to a subroutine of code and then returns
```
        GoSub Line1
```
`On .. GoSub` – On condition Branch to a specific subroutine then return at the next Return statement
```
        On Number GoSub Line1, Line2, Line3
```
* `On .. GoTo` – On condition branch to a specific line of code
```
        On Number GoTo Line1, Line2, Line3
```


Date & Time

`Date` – Gets the current date
`Time` – Gets the current time
`Now` – Gets the current date and time
`Timer` – Returns the number of seconds since midnight
`Year` – Returns the current year
`Month` – Returns the current month (integer)
`MonthName` – Returns the text of the name of a month
`Day` – Returns the current day
`Hour` – Returns the current hour
`Minute` – Returns the current minute
`Second` – Returns the current second
`WeekDay` – Returns the current day of the week (integer)
`WeekDayName` – Returns the text of a day of the week

Error handling

`On Error` - Enables an error-handling routine
`On Error GoTo Line2` (if error occurs, go to line2)
`On Error Resume Next` (if error occurs, continue executing next line of code)
`On Error Goto 0` (disables error handling)
`Resume` - Used to resume execution after a error-handling routine is finished
`Resume Next`
`Resume Line1`

This is just a very simple primer and we are going to stop there, because it should be enough to enable you to find your way around the IDE.

I would suggest that you go off and search the Internet for VB6 tutorials, there are thousands and if you look at and build a couple of simple examples you will quickly be hooked on programming.

Chapter 9

The Configuration File

To make it really easy on our beginner readers, we have written a configuration file, which should make life easier in regards to producing an XML file for SAPI Grammar.
SAPI requires an XML file in which is a list of words that SAPI will recognize. This list can be a straight list of Keywords or it can be in a tree form with branches to Sub keywords and commands. Now, to say SAPI is picky about its XML file is an understatement! We have touched upon the subject elsewhere in the book but here we are just saying that if you use our config file, which is written in plain old English, then the perl server script will take this config file and turn it into XML in the exact format that SAPI requires.

Take a look at the config file:

```
#
# Comment lines start with# and will be ignored
#
# KEY
# ---
# C - Command feature
# B - Board control  No Audio confirmation
# S - Script
# M - Multi/Macro
#
C|COMMAND_WORD|+Rachel
B|0|7|Light|Games Room|Dummy0
B|0|6|Light|Games Room|Dummy1
B|0|5|Light|Games Room|Pool Table Area
B|0|4|Light|Games Room|Arcade Area
B|0|3|Light|Games Room|Bottom of stairs
B|0|2|Light|Games Room|Pool table
B|0|1|Light|Bar|Bar Area
B|0|0|Light|Balcony|Balcony
B|1|7|Light|Bar|Spot above bar
B|1|6|Light|Bar|Spot above tv
B|1|5|Light|Bar|Spot by seating
B|1|4|Light|Bar|Spot by door
```

Chapter 9

```
B|1|3|Light|Bar|Dummy4
B|1|2|Light|Games Room|Alcove left
B|1|1|Light|Games Room|Dartboard
B|1|0|Light|Games Room|Alcove right
B|2|7|Light|Office|Office six
B|2|6|Light|Office|Office five
B|2|5|Light|Office|Office four
B|2|4|Light|Office|Office three
B|2|3|Light|Office|Office two
B|2|2|Light|Office|Office one
B|2|1|Light|Games Room|Entrance
B|2|0|Switch|Garden|Gates
S|9001|weather
S|9002|shares
S|9100|music=play
S|9101|music=pause
S|9102|music=stop
S|9103|music=prev
S|9104|music=next
S|9105|music=volumeup
S|9106|music=volumedown
S|9107|music=repeat&enable=1
S|9108|music=repeat&enable=0
S|9109|music=getplaylisttitle
M|801|All lights
M|802|Office Lights
```

Let's break it down and see what we have.

The # commented lines are ignored by the parsing script in the perl server program, and they are there just to show you how commands are formatted with their Identifier letter.

You will see that the "C" identifier denotes the initial command word "Rachel". This is the Keyword that Rachel sits there listening for, and "she" should ignore every other word spoken. Of course, after hearing "Rachel" she will listen for the next few words to try and build a hypothesis.

The identifier "B" denotes a Board function. That is, we are instructing the interface board to carry out a command, such as switch a relay on or off.

The identifier "S" denotes run a script, such as the weather script where Rachel goes away to the internet, gets the weather and reads it back.

Finally the identifier "M" denotes a Multi/macro, which is reserved for items like "All lights on".

The Configuration File

When we look at a line in the config file for Board Control commands (identifier B) we see the identifier followed by a number between 0-3. This number represents the board address we have assigned to the board on which we want to carry out the relay command. In the next column we have a number between 0 and 7; this number represents the output channel (relay) number on the board.

Next column is an item identifier such as "light". This enables us to group items together logically so that if we wanted to turn all lights on or off we just have to search for the identifier "light". Next column is the location, as in "Bar", and the final column is the item name such as "Bar area".

An example of a "B" identifier line is: `B|0|1|Light|Bar|Bar Area`
Decoded, this means, Board command, on Board number 0, Output Channel (relay) 1, type is light, location is Bar and the item name is Bar area.

Looking at an identifier "S", the lines are different. We start with the "S" identifier then go straight to a four digit code such as 9001 and in the next column is the script name, in this case "weather".

On the "M" identifier, we have a three digit code beginning 801 which denotes "All lights"
Our perl server script uses these codes to build a perfect XML file and places it in the correct path for the SAPI engine to find.

Chapter 10

The Rachel Program

In this chapter we are going to go thoroughly through the programming in VB6 of the Rachel program. What are we trying to achieve?

> Yes you could write this in twenty other programming languages, but this is the easiest and it works very well.
>
> Later in the book we will show you how to do some Perl scripts, which are not so visual.

Well, at the end of this brief explanation we will show you a flow diagram of sorts which will outline the basic functions that we want to achieve.
But first let me talk you through what's going on when we run Rachel.
First off, the Graphical user interface (GUI) appears, that gives you a clue that something is happening. The program then is made to wait whilst it calls for the Perl server to start. When the server has started, the program searches for the grammar xml file.
It's lucky we started the perl server script, because that script is responsible for creating the grammar file from our human-readable config file. Having loaded the grammar, the program sets about creating a SAPI object to enable it to access the SAPI engine. If the program fails to load SAPI it will stop with an error message.
Upon successful loading of SAPI, the program then sets about starting the Socket connection with the perl server program.
If the connection fails the program will halt with an error message; if it succeeds then it sends off an information request to the perl server.
The information request causes the perl server to query its config file and send back a list of names that it finds, such as Garden lights, office light etc. The VB6 program then uses this returned information to populate the four text boxes that represent the various channels of the interface board.

We then enable the SAPI engine to allow it to start work. We immediately also enable the microphone connection so that SAPI has something to listen to. The program then looks to see if we have requested an event information to be displayed. If we have, then it starts to display that information in the event display text box.

Now it starts the hard work, the program sits and listens to everything it hears, waiting patiently for a Keyword, in this case it is "Rachel". If it hears the keyword and is confident about the level of recognition then it will listen out for a phrase or command word that is contained in the grammar file. If it correctly identifies a command it will send a code string via windsock to the perl server consisting of a number of digits and a request for status. These digits represent the interface board or script that we want to activate. The perl server interprets these strings and acts upon them, the status request cause the perl server to query its status file to see what is switched on and what is not. The perl server then responds with the status and sends back to the VB program a confirmation that the command has been received and acted upon, along with the current status of the interface board. The VB program then displays the returned confirmation data and updates the led lights of the interface boards on the GUI . The program then sits and awaits the next recognition of a keyword.

Now that's given you half of the picture as to what is happening. The entire picture will become clear when you read the functionality of the perl server script in Chapter 11.

And why aren't we using perl to do the talking, and switching?

Well, we want to show you that you can do it in VB6, perl, mix and match, whatever. But this method we are showing you gives you, the reader, the best balance and it makes it all very user friendly. To help understand the flow of the program we've produced an elementary flow diagram which shows the critical paths and ultimate resolution.

The Rachel Program

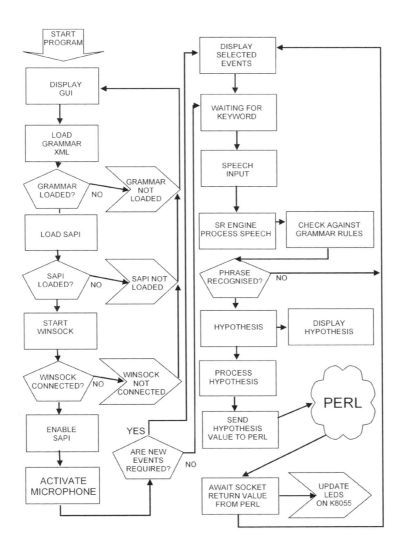

Chapter 10

Now you could, and I am sure lots of you will, skip to the downloaded software from the www.elektor.com support page, load the complete program onto your PC and start playing with it, but you're missing out. By doing it yourself, you will understand how and why it works and then you will be capable of modifying the program to meet your individual specific needs. So try and contain yourself!!

This is what we are going to create:

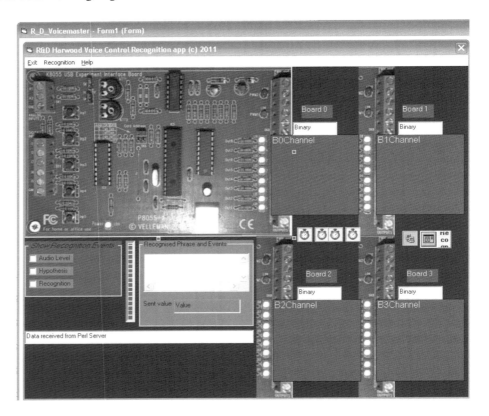

To be sure it's a masterpiece, all our own work, please give generously! Ha.

This is our VB6 Voice Recognition GUI interface using the Velleman K8055 and K8056 interfaces. Later in the book we will briefly cover the PC-Control interfaces and the interface GUI which displays all the information we require to operate and control the system.

The Rachel Program

Just a quick explanation of what the graphic represents. We show top left an image of a whole k8055 board, with superimposed led graphics over layed light up at appropriate times. To the right of the k8055 board image is a series of four blue panels, each representing another k8055 interface board. They also have led graphics superimposed, and above them is a small text panel that displays a binary value of the channels for that board. The Red panel is the events pane. It shows real time events that we have selected from the checkboxes to the left in the blue panel. The text panel at the bottom of the red panel displays the code that it sends via winsock when recognition is achieved. The whole red panel turns green for a few seconds each time the program recognises a word or phrase. Sandwiched in between is a blue, tall progress bar which acts as our VU meter and finally, below that, is a text panel that displays all the messages sent from the perl server and received by the winsock control in our program

Let's start.

Open the VB6 IDE and select a new Exe project. You have seen the IDE earlier in the book so from now on I will only show smaller graphics of relevance, (try and save some trees!!).

You will need to open a form.

Select Properties and change the following:
- **Height**8325 and **Width**8340
- **BackColour**&H00000080& (this sets it to a dark maroon but you can select any colour you want)
- **Caption**   Set Caption to whatever you want to appear along the top. The only thing you can't do is place copyright on it because, if you are following these instructions then you are using our code and you don't own the copyright, we do. Ha Ha!

Your properties, apart from the Caption, will look like this:-

Chapter 10

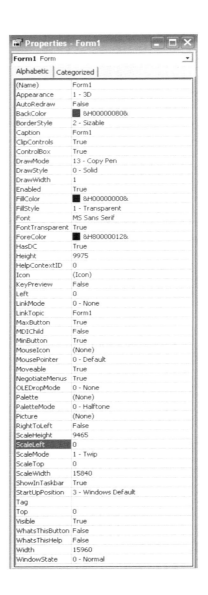

You will need to save your project, so at the top of the IDE , Select File, Menu, Save as , and in the Dialogue box select whether to accept our default project name Rachel or choose your own. By saving the project you will also have saved Form1. You will get sick of me saying this, but save your work every time you do any amount of coding. Make it a habit. ***HIT SAVE PROJECT! You are going to see this a lot!***

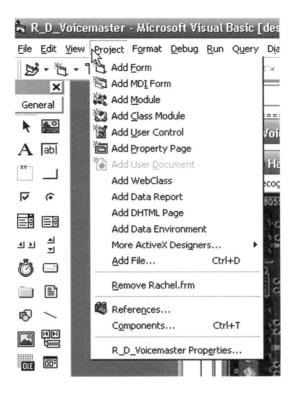

There is nothing worse than running a part built application that crashes and loses the IDE and all your hard work, if you save your program regularly then if the IDE crashes you won't have lost much if anything. Lecture over!

This will cause the above box to appear. You must check down the list and ensure that you have all the same items selected as shown, and moved to the top of the list, as windows works on a priority system and our selections must be near the top for correct execution.
We now need to add some outside references.

Chapter 10

In a previous chapter we said we needed to download and install SAPI 5.1 sdk and this must be done before we go any further, as we are going to make reference to speech libraries, so if you haven't done it yet, do it now!

To add the references we select Project menu and References

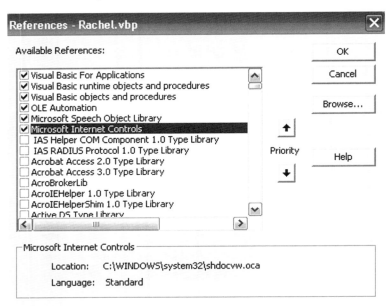

These are the must haves:
- **Microsoft Speech Object Library**
- **Microsoft Internet Controls**
- **OLE Automation**
- **Visual Basic objects and procedures**
- **Visual Basic runtime objects and procedures**
- **Visual basic for Applications**

Then click OK to add to our project.

Chapter 10

These library controls allow us access to windows speech recognition and it is these that we call to get our recognition done. We are now going to add some extra components that we will use in the project, the components live down the left hand side of the IDE, and you will see a number already installed by default.
Right click on the components panel and select Components, this will display this box.

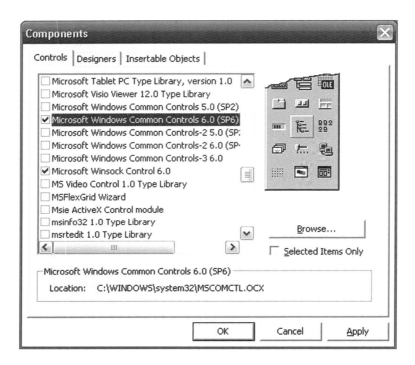

You must add these components:
- Microsoft Winsock Control 6.0
- Microsoft Windows Common Controls 6.0 (SP6)
- Microsoft Common Dialogue Control 6(SP6)

Click Apply and OK, the components will then appear in the component selector panel.
We use the Winsock control to "talk" to Perl scripts and the Common Controls to display our simple VU meter (sound level).

We will now drag some components onto our form, it's a bit of a list, but doesn't take long.

We will set the properties of each a little later on, just get them on the page first, and drag them anywhere. We can set then up later.

The Rachel Program

HIT SAVE PROJECT!

Drag a **Winsock control** to the form (technically it's not a drag it's a click, select then click on form and drag to desired size). Then drag these components to the form:
- 4 x Timer components
- 3 x Checkbox components
- 2 x Frame components
- 7 x Text Box components
- 1 x Progress bar (drag it downwards on your form then it will go to a vertical progress bar) component
- 1 x Command Button component
- 1 x Common Dialogue Box component
- 37 x Image components
- 4 x Label components

Your form will look a mess, something like this!

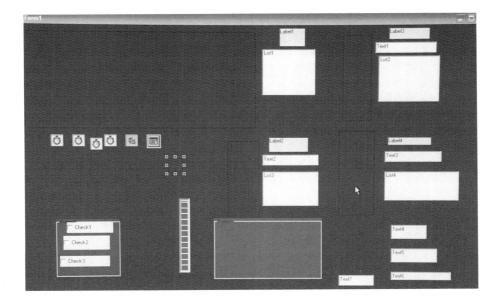

Not a problem, let's lay it out roughly as we want it to look, remember some components are visible in Runtime and some are not. Drag the components around and leave a space at the bottom left of the screen. Drag **Frame2** into the corner and expand it until it is about the size shown.

Chapter 10

In Frame2, Properties:
- Select Caption and add "Show Recognition Events" without the quotes, then select font and set text size to 12 pt, bold italic.
- Select BackColor and set to &H00FF0000& (which is vbBlue)
- Right click on Frame2 and select Bring to front. This ensures we can always see it on the form.

HIT SAVE PROJECT!

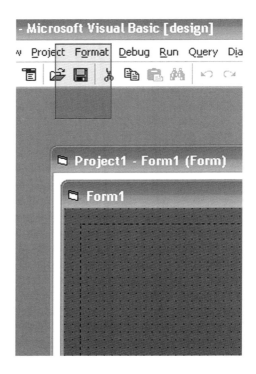

Good, we are making progress!!

Now we will populate Frame2 with three CheckBoxes, but before you do this right click them one at a time and select Bring to front, else you will find they disappear!

We are going to alter the properties on each of these three CheckBoxes so select them all by holding down the shift key whilst clicking each one then in properties. Select BackColor and change to &H000000FF& (which is vbRed). Then click elsewhere on the form to release the group select.

The Rachel Program

Next we are going to give them more meaningful names, so alter the Properties thus:

	Check Box		Properties
Old Name	New name	Caption	Value
Check1	AudioLevel	Audio Level	unchecked
Check2	Hypothesis	Hypothesis	unchecked
Check3	Recognition	Recognition	unchecked

Then position them like this:

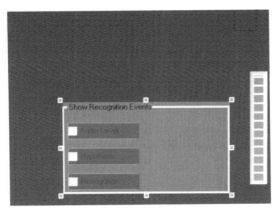

HIT SAVE PROJECT!

Next we turn our attention to Frame1:
- Drag other components out of the way and clear an area bottom right of the form. Drag Frame1 corner so that it measures Height 169, Width 265.
- In Properties change color to&H000000FF&vbRed.
- Change its Caption to "Recognised Phrase and Events"
- Drag TextBox5 onto Frame1 and change its size as shown, position it centrally, but towards the top. In Properties change Scrollbars to show number 3-Both.Change its name to EventText-Field
- Drag Label5 and place it on Frame1 bottom left, change its properties to backColor&H000000 FF&vbRed then Caption to Sent value
- Drag TextBox6 to Frame1 bottom right and change its Properties to Backcolor&H000000FF& vbRed and Text to "Value".

Chapter 10

Your finished frame 1 should look like this:

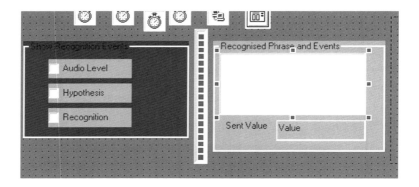

As you can see, I have dragged the ProgressBar1 alongside and changed it's Height &Width to look as shown, ProgressBar1 is going to be our VU meter(sound level).

HIT SAVE PROJECT!

Now, we are using the Velleman USB Experimenter Board (K8055) or (VM110) in this project . It's nice to have a graphical representation of what's going on, so we added a background graphic and some foreground graphics which we will bring into play later on.

The graphics can be found on the www.elektor.com support page as ak8055.jpgaddboard.jpg
- Select image1
- Change Properties to Stretch = true
- And set picture to ak8055
- Then select image 2, 3 and 4
- Change their properties to stretch = true and set the picture to addboard

Your board should look like this:

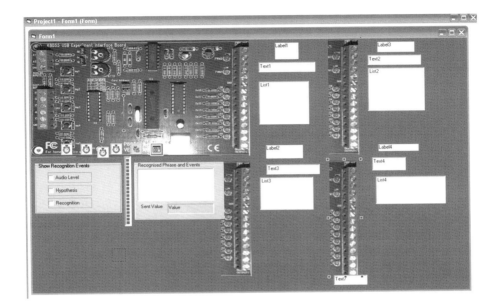

Now we are going to add the lit leds. Select another image5 component on the form.
In its Properties select:
- Picture and Navigate to led.jpg
- Name   poweron
- Visible   False
- Stretch   True
- Height
- Width (values vary according to screen resolution, just make them big enough to cover the leds in the images)

Right click the image and select bring to front, else it will disappear behind some other component.

Drag this led and place it on top of the power led in the main k8055 image.

Create another image component and place it on your form, Change its properties to
- Name   led
- Stretch =true
- Picture   led
- Resize as before

Chapter 10

Right click on the led image and select copy then paste onto the form. You will be challenged by the IDE asking if you want to create an array, select yes and paste 31 more leds to the form, that's 32 total in the array.

Now the tedious bit, starting with led (1) place them on the form over the leds starting with (1) on bottom of the main board as shown.

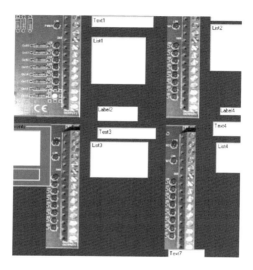

HIT SAVE PROJECT!

Led(9) goes on the next board to the right at the bottom, thus Your form should now look like this..

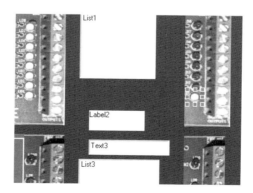

Populate the leds then when you get to number 17 position it thus:

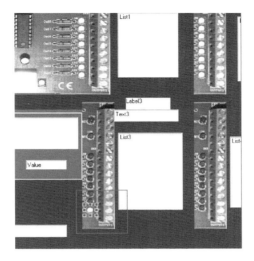

HIT SAVE PROJECT!

Continue populating up until led 25 which goes bottom right of the fouth board, continue until your board looks like this:

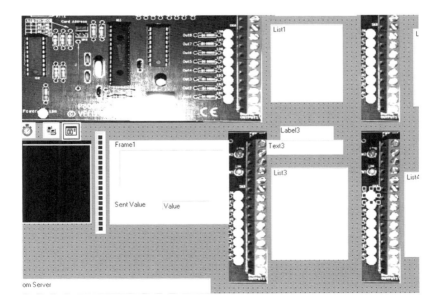

Chapter 10

You're beginning to get an idea of what the finished program will look like.
Now we are going to add some menu items. Select the Menu editor from the top of the IDE

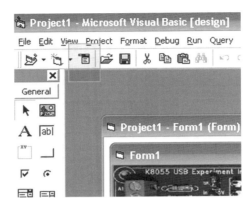

This brings up our dropdown menu editor. The menu editor allows us to place a structured menu at the top of our form, just like you see in any other windows application.

We are going to add the following menus:
- In the Caption box type "File" (don't type the quotes)
- In the Name box type "File"

Click next then type in the Caption box "Recognition" and the same in the name box, click Next, and
Do the same using the word "Help".
This will produce the three top level menu items; we can now add the Sub menu items.

The Rachel Program

In the list click Recognition, to highlight it, then click Insert, Then Type "Exit" into the Caption and Name boxes, (if you want to create a keyboard shortcut to any of these menu items just place an "&" in front of the letter you want to use as a keyboard shortcut), then click the arrow pointing right, this will place the word Exit into a sub menu of File, your menu box will look like this:

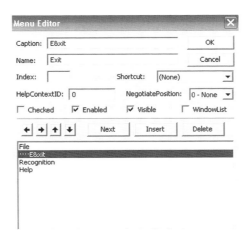

We will now, using the same procedure add a sub menu item to the Recognition menu item, add "Start Recognition" and "End Recognition" to the Recognition menu and then add "Show Grammar" and "About" to the Help menu item. So your Menu tree looks like this:

83

Chapter 10

HIT SAVE PROJECT!

That's the Menu finished. Next we need to select labels 1 to 4.
- Set Properties backcolor = &H00FF0000& (vbBlue)
- Then Set their captions to Board 0 to Board 3 as shown.
- Next text 1 to text4, we need to change the Properties
- Text to the word Binary
- Name to Binary0 to Binary 3

so they look like this:

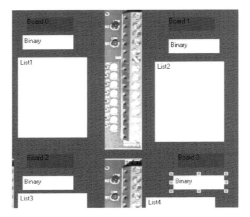

Next we turn our attention to listbox1 through to listbox4.

Listbox1 properties

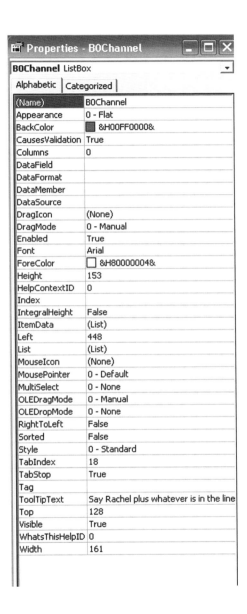

For listboxs 2 through 4 change the name to B1Channel, B2Channel and B3Channel respectively.

Your listboxs will look like this:

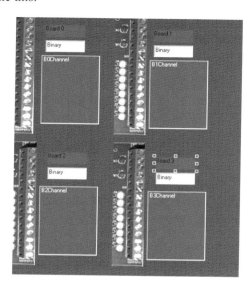

Last item is Text7. We need to drag it and extend it to look like this.

Change its Text property to "Data Received from Perl server". and its Name to DataIn.
Great! Done! Your finished form should look like this, unless you have made other, deliberate changes of course.

The Rachel Program

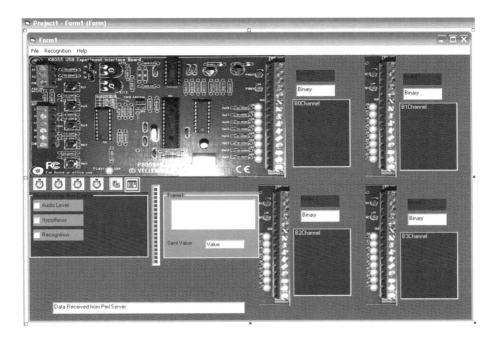

All we need to do now is tie all the items together and add some functionality.
Believe it or not, and if it's no then try it yourself, we have a working program right now! When I say working, I mean that all the components are working and they display and exhibit standard windows behavior. If you click the Run triangle at the top of the IDE, your program will compile and run.

It won't do anything speech recognition wise but it will run and all the click boxes etc. will function. You're a programmer already!

Ok, I am going to drop the ***HIT SAVE PROJECT!***

So from now on, you are on your own, in that respect, you've been warned, ***SAVE regularly. Nag! Nag! Nag!*** Back to the job in hand, we are now going to add functionality.

Chapter 10

VB6 Code adding the functionality

To do this we double click on our form and it will open a code window. It is here that we are going to add the main part of the functionality.
I shall show you some coding, then attempt to describe what it does. *Sometimes I look at my code and think, who wrote that ?Or as Darren would say, "what wrote that!". He can be very hurtful.*
First we must start with some Declarations and Constants. On Project 1, form1 code page, select General from the left hand drop box and Declarations on the right. We then type in the following "verbatim". On any line everything after a ' is treated as a comment. I have shown these comments in light text. You do not need them for the program to work, but you do need them to remember what does what.

This code is case sensitive. You must spell and type characters exactly as shown.

```
'===============================================================================
'
' This code is © R&DHarwood 2011. No part of this code may be copied, or  'used
unless  a copy of the book PC Voice Control System by R&D Harwood is  'owned by
that user, and this notice is preserved in the coding.
'
'===============================================================================

Option Explicit

    Public WithEvents RC As SpSharedRecoContext    'The main shared Recognizer Context
    Public Grammar As ISpeechRecoGrammar           'Command and Control interface
    Public board0binary As String                  ' Binary value holder for board 0
    Public board1binary As String                  ' Binary value holder for board 1
    Public board2binary As String                  ' Binary value holder for board 2
    Public board3binary As String                  ' Binary value holder for board 3
    Public board0 As Long
    Public board1 As Long
    Public board2 As Long
    Public board3 As Long
        Dim indent As Integer                      'Sets indent level for small
output window
        Dim fRecoEnabled As Boolean                'Is recognition enabled
        Dim fGrammarLoaded As Boolean              'Is a grammar loaded
        Dim RecoResult As ISpeechRecoResult        'Recognition result interface
        Dim grammy As String
        Dim info As String                         ' Holder for value returned by
```

VB6 Code adding the functionality

```
perl server
        Const WM_USER = &H400                          ' VU meter colours
        Const CCM_FIRST = &H2000&
        Const CCM_SETBKCOLOR = (CCM_FIRST + 1)
        Const PBM_SETBKCOLOR = CCM_SETBKCOLOR
        Const PBM_SETBARCOLOR = (WM_USER + 9)

Private Declare Sub Sleep Lib "kernel32" (ByVal dwMilliseconds As Long) ' required
for sleep call
Private Declare Function SendMessage Lib "user32" Alias "SendMessageA" _
(ByVal hwnd As Long, ByVal wMsg As Long, ByVal wParam As Long, lParam As Any) As
Long ' required to send messages internally
```

In this section of code we have declared some variables that we want all of the program to be able to see. We have also declared some constants and a couple of system calls to Kernel32 and user32, remember anything in light text is a comment and is not required for the program to function.

Now the code in VB6 is not in any particular order. The program is event driven and the flow of the program is determined by which particular events get fired and responded to. So what I am saying is that the code does not flow entirely logically; it helps to think of it as a collection of subroutines that can be in any order that get called by events. Clear as mud?

Let's look at the next section of code.

```
Private Sub B0Channel_Click()
toggler B0Channel, board0binary, 0 ' switch any clicked item
End Sub
Private Sub B1Channel_Click()
toggler B1Channel, board1binary, 1 ' switch any clicked item
End Sub
Private Sub B2Channel_Click()
toggler B2Channel, board2binary, 2 ' switch any clicked item
End Sub
Private Sub B3Channel_Click()
toggler B3Channel, board3binary, 3 ' switch any clicked item
End Sub
```

This snippet of code follows right on from the last, and deals with receiving the clicks on any board to toggle the on or off state, by calling the toggler sub routine. You will notice there is a click event for each board 0 to 3. The next bit of code is the toggler sub routine.

Chapter 10

```
Private Sub toggler(kboard As ListBox, boardbinary As String, c As Integer)

Dim int0 As Integer  ' Define counter integer
Dim send0 As String  ' Define the winsock string to send for the board
Dim bit0 As String   ' bit holder
Dim bit01 As Integer ' bit test holder
Dim boardbinaryrev As String 'Define the binary string reverse holder
   With kboard  ' use Board 0 Channels
      For int0 = 0 To .ListCount - 1 ' set counter limits
         If .Selected(int0) Then 'cycle through all list see whats selected
            boardbinaryrev = StrReverse(boardbinary)  ' reverse the binary string
            bit0 = Mid$(boardbinaryrev, 8 - int0, 1)  ' find the bit we are trying
to set
            bit01 = InStr(bit0, "1")  ' assign bit to varaible for testing
            If bit01 = 1 Then ' test bit
               send0 = "1" & c & 7 - int0 & "0"  ' if bit set toggle to off
            Else
               send0 = "1" & c & 7 - int0 & "1"  ' if bit not set toggle to on
            End If
               Text1.Text = send0 ' debug show data
                 Winsock1.SendData send0 ' send the command
         End If
      Next
   End With
End Sub
```

The toggle Sub Routine takes two arguments, or parameter's; the binary value for the board and the board number. It looks at the selected board and see's what channels are on or off and depending on the existing value sets the channel to the other state. I.e. if channel 1 is on; toggler sets it to off. It does this by sending a message through winsock to the Perl Server which then carries out the command. Now this is where we appear to be getting out of flow order, since this Routine deals with events and settings when the form first loads.

```
Private Sub Form_Load()
Shell """C:\launchdch-serv.bat"""  ' launch a batch file to start the Perl Server,
it must start before this program
Sleep (1000) ' wait for Perl Server to start

    On Error GoTo Err_SAPILoad     '   Set up error handler
       Timer2.Enabled = True   ' send status command to Perl Server to load k8055
```

VB6 Code adding the functionality

```
listboxes
    Winsock1.Connect  ' connect to Perl Server
    Poweron.Visible = True  ' show board power on led
      '   Initialize globals
    indent = 0
    fRecoEnabled = False ' dont start the recognition just yet
    fGrammarLoaded = False ' dont load grammar yet
      '   Create the Shared Recognition Context by default
    Set RC = New SpSharedRecoContext
    InitEventInterestCheckBoxes 'initialise the event checkboxes
      '   Create the grammar object
    LoadGrammarObj
      '   Attempt to load the default rh2.xml file
    LoadDefaultCnCGrammar

   Call Recognition_Click ' call recognition sub routine
    Exit Sub
Err_SAPILoad: ' error handler, note any fault will call this on startup
    MsgBox "Error loading SAPI objects! Please make sure SAPI5.1 is correctly
installed.", vbCritical
    Exit_Click
    Exit Sub
    End Sub
```

The first thing the form does as it loads is to make a Shell call which starts a batch file called launchdch-serv.bat.

Launchdch-serv.bat is a tiny batch file that runs in a command window, in the background. Its sole purpose in life is to start the Perl Server running. The batch file looks like this:

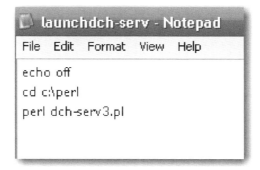

91

Chapter 10

Hardly anything there, eh? Line one just tell the command prompt not to show the commands. Line two says to change the working directory to c:\perl, which is where all our perl scripts live, and the third line says to use perl to run the script named dch-serv3.pl, which is our perl server script.

This script has to be run prior to anything else happening in the VB6 program.In order to ensure that the perl server gets up and running we make the VB6 program sleep for 1 second. When it awakes, the Perl server is running thus allowing the Winsock1.connect event to happen without returning an error, which it would otherwise do if there was no Perl server to respond to the winsock connection request.

We then go on and turn poweron led graphic to visible, which lights up the power led in our K8055 board graphic. At this point we must make sure that the SAPI5 has not loaded any grammar or started recognition before we are ready for it; so we set both events to false.

> Winsock is not very forgiving, if it makes an attempt to connect and the connection is refused for any reason, it returns an error. So it is vital that we make sure the server is running before we attempt to connect to it.

We then look to see what event we want shown in our event view window and now we are ready to load the grammar.

Now, any errors that occur during the form load will call the Err_SAPILoad error handler. This is just a general handler that brings up a message "Error loading SAPI objects". This is a generic message, and will be generated by whatever caused the error at that point. So when you are debugging, just remember that the message is generic and the fault could, in fact, be caused by any instruction failing in that sub routine.

The next snippet deals with loading the grammar and starting the recognition Engine.

```
Private Sub LoadGrammarObj()
    Set Grammar = RC.CreateGrammar(1)  ' set the grammar
End Sub
Private Sub LoadDefaultCnCGrammar()
    '    First Grammar load attempt
    On Error GoTo Err_CFGLoad  ' in event of error use handler
    Grammar.CmdLoadFromFile "c:\rh2.xml", SLODynamic
    '    Set rule state to inactive until user clicks Recognition button
    Grammar.CmdSetRuleIdState 0, SGDSInactive
```

VB6 Code adding the functionality

```
    '   Set the Label to indictate which .xml file is loaded.
    grammy = "c:rh2.xml" ' debug label
    fGrammarLoaded = True ' initialsie the grammar
    Exit Sub

Err_CFGLoad: ' Error handler
    Frame1.Caption = "No Grammar file loaded can't continue"
    Exit Sub
End Sub
Private Sub Recognition_Click() ' this isnt really a command button click, it is a
sub pure and simple
    On Error GoTo ErrorHandle ' error handler, will be envoked on any error
    Grammar.CmdSetRuleIdState 0, SGDSActive ' start the recognition process
           Recognition.Caption = "Start Recognition" 'show recognition in
progress
           Poweron.Visible = True ' show power led on
       Exit Sub

ErrorHandle: ' error handler for this sub routine
    MsgBox "Failed to activate the grammar. have you moved c:rh2.xml?", vbOKOnly
End Sub
```

The Sub LoadGrammarObj() tells VB6 to create a grammar object.

The Sub LoadDefaultCnCGrammr() starts off with an error handler. All Subs should be able to handle their own errors. This is really useful when developing and debugging. The message is generic again and will be shown for any error that occurs in that sub routine. In this particular case, the message states that no grammar file has been loaded and therefore the program cannot continue - which is correct.

Next, our grammar file is called; that's the c:\rh2.xml file, remember this file is generated from our config file by the Perl server program. The xml file is dynamic, in so far, as whenever you start the Perl server it looks at your config file and generates a new xml - based on what it sees in the config file.

We then initialise the grammar in order to make it available to the Voice recognition Engine.

The Sub Recognition_click() may appear to be a command button click event in the listing, but it is not; we do not have a command button. We chose to write the sub this way to show that a sub is a sub is a sub..., that is to say, no matter what its label, you can call it by name or event.

Chapter 10

> If the grammar fails to load, it does not stop our program, but it does stop the SAPI engine from attempting to recognise anything, so we must have the correct grammar in the correct place on startup.

A Recognition click makes the grammar active and starts the recognition Engine. Should it fail to do so, the error is handled by the error event handler.

The next block of code deals with getting and displaying the events we want to monitor such as audio levels, recognition, hypothesis etc. The Sub UpdateEventList is a simple piece of code to display the events we want to see in the view window, and in order to make the data clear it indents the data according to type.

```
Private Sub UpdateEventList(StreamNum As Long, StreamPos As Variant, szEvent As
String, szEventInfo As String)
    Dim szIndent As String  ' setting up the event viewer
    Dim szStreamInfo As String
    Dim i As Integer

For i = 0 To indent - 1
        szIndent = szIndent & "    "
Next

    EventTextField.Text = EventTextField.Text & szIndent & szEvent & szStreamInfo
& szEventInfo & vbCrLf  ' showing events
    EventTextField.SelStart = Len(EventTextField)
    EventTextField.SelLength = 0
    Timer1.Enabled = True
End Sub


'   The following subroutines are event handlers that get called when the SR
engine
'       fires events.
'       Audio Level event handler
Private Sub RC_AudioLevel(ByVal StreamNumber As Long, ByVal StreamPosition As
Variant, ByVal AudioLevel As Long)
    UpdateEventList StreamNumber, StreamPosition, "AudioLevel", " [Level=" &
AudioLevel & "]"
    Dim audi As Long
    'ChangePBForeColour ProgressBar1.hwnd, vbRed
```

VB6 Code adding the functionality

```
        audi = AudioLevel
If audi < 25 Then ' poor volume level
    ChangePBForeColour ProgressBar1.hwnd, vbRed
End If
    If audi > 25 And audi < 70 Then ' good volume level
        ChangePBForeColour ProgressBar1.hwnd, vbGreen
    End If

If audi > 70 Then ' too much volume
        ChangePBForeColour ProgressBar1.hwnd, vbRed
    End If
    ProgressBar1.Value = audi
End Sub
```

The Sub AudioLevel grabs the audio level from the windows system and places it in the variable named audi. Since we are using this data to drive our vertical VU meter, we must process the value to see if it exceeds certain thresholds. If it does we change the colour of the meter; Red for too low or too high and Green for correct level.

The last event we look at is the Hypothesis. This is the code that fires when a Hypothesis is achieved.

```
'   Hypothesis event handler
Private Sub RC_Hypothesis(ByVal StreamNumber As Long, ByVal StreamPosition As
Variant, ByVal Result As SpeechLib.ISpeechRecoResult)
    UpdateEventList StreamNumber, StreamPosition, "Hypothesis", " [Text=" &
Result.PhraseInfo.GetText() & "]"
End Sub
```

It places the resultant hypothesis in the events text area.

We now address the recognition Event with the Sub RC_Recognition.

```
' Recognition achieved
'   Recognition event handler
Private Sub RC_Recognition(ByVal StreamNumber As Long, ByVal StreamPosition As
Variant, ByVal RecognitionType As SpeechLib.SpeechRecognitionType, ByVal Result As
SpeechLib.ISpeechRecoResult)
    Dim RecoNode As Node
    Dim sendatGroup As Integer ' setting variables to hold data to be sent to
```

```
perl server
Dim sendat As String
    Dim sendat1 As String
    Dim switch3 As String
    Static i As Integer
'   Update Event List window first
     UpdateEventList StreamNumber, StreamPosition, "Recognition", " [Text=" & Result.PhraseInfo.GetText() & ", RecoType=" & RecognitionType & "]"
     Timer1.Enabled = True
'   Increment unique value for RecoNode's key name.
     i = i + 1
'     Save the recognition Result to the global RecoResult
    Set RecoResult = Result
     Frame1.BackColor = &HFF00&
     Timer1.Enabled = True                        ' enable timer to lock out recogintion until this current recognition is complete
     sendat1 = Result.PhraseInfo.GetText()        ' gets the recognised phrase places into sendat1 string
     switch3 = Result.PhraseInfo.Properties(0).Value ' gets the recognised phrase value set in gramma rh2.xml
     Frame1.Caption = sendat1                     'displays recognised phrase
     Text1.Text = switch3                         'displays recognised phrase value
    If Winsock1.State = sckConnected Then         ' check if winsocket is connected
     Winsock1.SendData (switch3)                  ' send the phrase value to perl winsock server for processing
   Else
     Text1.Text = "No socket connection "         ' issue warning no connection equal no command responded to
   End If
   'end send data to darrens perl stuff
     '  end recognition achieved

End Sub
```

When the Recognition event fires this sub is called. It gets the Recognition event and pushes it to the events window. It then fires Timer1, which changes the colour of the event display frame to green to show that the event has occurred, and then clears the text in the event window. Finally, it changes the frame back to its original red colour.

VB6 Code adding the functionality

The Phrase recognised is sent to the events frame caption for display.
The Value of the recognised phrase is sent to the Perl Server via windsock1 so that the Perl Server can take the appropriate action. There is no error handler in this sub, but should the winsock fail it will drop through to a "no socket connection" message.

In the next block of code, we first look for change of state of the Recogniser Engine with the Sub RC_RecogniserStateChange

```vb
'    Recognizer State Change event handler
Private Sub RC_RecognizerStateChange(ByVal StreamNumber As Long, ByVal
StreamPosition As Variant, ByVal NewState As SpeechLib.SpeechRecognizerState)
    UpdateEventList StreamNumber, StreamPosition, "RecognitionStateChange", "
[NewState=" & NewState & "]"
End Sub
'    Stream Start event handler
Private Sub RC_StartStream(ByVal StreamNumber As Long, ByVal StreamPosition As
Variant)
    indent = 0
    UpdateEventList StreamNumber, StreamPosition, "StartStream", ""
    indent = 1
End Sub
'    The following subroutine enables the event interest check boxes based on the
selected boxes
Private Sub InitEventInterestCheckBoxes()

    If RC.EventInterests And SREAudioLevel Then ' reset values
       AudioLevel.Value = Checked
    End If
      If RC.EventInterests And SREHypothesis Then ' reset values
       Hypothesis.Value = Checked
      End If
        If RC.EventInterests And SRERecognition Then ' reset values
          Reco.Value = Checked
        End If

    End Sub
Private Sub SetEventInterest(EventInterest As SpeechRecoEvents, EventCheckBox As
CheckBox)
    If EventCheckBox.Value = Checked Then     ' use selected events
       RC.EventInterests = RC.EventInterests Or EventInterest
```

```
    Else
 RC.EventInterests = RC.EventInterests And Not EventInterest
    End If
End Sub
'    Audio Level event interest  debug purposes
Private Sub AudioLevel_Click()
    SetEventInterest SREAudioLevel, AudioLevel ' set audio events for audio
End Sub
'    Hypothesis event interest debug purposes
Private Sub Hypothesis_Click()
    SetEventInterest SREHypothesis, Hypothesis ' set Hypothesis events
End Sub
'    Recognition event interest debug purposes
Private Sub Reco_Click()
    SetEventInterest SRERecognition, Reco ' set recognition events
End Sub
```

The Subs, RC_StartStream, InitEventInterestCheckBoxes, SetEventInterest, Audio level_Click, Hypothesis_Click and Reco_Click deal with viewing the events that we have previously checked the boxes to view. Those subs simply look to see if the checkboxes have been ticked and if they have adds the event to the eventinterest queue for display in the events of Interest frame.

We now move on to the menu items and what happens when we click a menu item.

```
Private Sub ShowGrammar_Click()
  Poweron.Visible = False   ' turn off poweron led
  MsgBox "Grammar File loaded is :- " + grammy, vbInformation, "About RecoVB" '
menu about grammar
End Sub
'    About box
Private Sub About_Click()
  MsgBox "(c) 2011 R&D Harwood. All rights reserved.", vbInformation, "About
Rachel" ' fantastic authors blurb
    End Sub
' Graceful exit
Private Sub Exit_Click()
    Unload Form1 ' relase form and associated memory, close program
End Sub
Private Sub startrecognition_Click()
  Call Recognition_Click ' start Recognition process
  Poweron.Visible = True ' indicate throgh power on led that recognition is active
End Sub
```

```
Private Sub stoprecognition_Click()
   Grammar.CmdSetRuleIdState 0, SGDSInactive ' tun off the grammar to cease
Recognition
   Poweron.Visible = False ' turn off power on led to indicate system not
recognising
End Sub
```

I think that they are mostly self explanatory, with an example of the stop recognition_Click sub. We do this by making the loaded grammar inactive. Now, without the grammar the program cannot run, but it keeps the Recognition Engine in an active state so you should think of this as a pause button rather than a permanent stop. A permanent stop would be implemented by using the exit menu, where the form is unloaded and memory released back into the operating system pool.

The next four blocks of code deal with the events that are fired when the timers are called.

```
' shows recognition box colour change red = normal listening, green = phrase
recognised
Private Sub Timer1_Timer() ' when Recognition achieved frame changes colour to
green for a short period of time
   Frame1.BackColor = &HFF&
   EventTextField.Text = "" ' clears the event field of extraneous data
   Timer1.Enabled = False ' governs the length of time that the frame stays green
End Sub
Private Sub Timer2_Timer()
'On Error GoTo ErrorHandle
   Winsock1.SendData "status"  ' get the status from the status file, so the system
starts at the last know setup
   Timer2.Enabled = False
   Timer3.Enabled = True ' wait a tad for data to be received then get info from
perl server
End Sub
Private Sub Timer3_Timer()
   Winsock1.SendData "info" ' ask Perl server for current info
   AudioLevel.Value = vbChecked ' autostart the audio level events display
   Timer3.Enabled = False    ' turn off timer 3
End Sub
Private Sub Timer4_Timer()
   Timer4.Enabled = False ' used to insert a short delay whilst valid data returns
from Perl server
End Sub
```

Chapter 10

- Sub Timer1 changes the colour of the events frame to green when a recognition event occurs, it then changes the color back to red after a short 1 second delay,
- Sub Timer2_Timer gets called when we request the status from the Perl Sever. It sends a winsock request to the Perl Server for the current status of the switched channels, and then calls timer3 in order to create a delay long enough for the Perl Server to respond with the current status data.
- Sub Timer3_Timer waits for a second then sends off a winsock request for the current info status from the Perl Server. It also calls the audiolevel checkbox and sets it to checked so that we start to get visual feedback that the system is live.
- Sub Timer4_Timer is a delay timer called from various places within the program in order to cause a delay whilst data is retrieved from the Perl Server.

As a matter of interest the timer doesn't halt the entire program- in fact only the sleep command will do that . What it does is to introduce an extra delay into the program stream.

Next we look at where a lot of the action happens; when Winsock1 recieves data the dataarrival event is triggered

In the listing directly after the Dim's we intercept the winsock data and place it in a variable returndata. Now, the data arriving into winsock, can be from different sources, so we must first determine what type of data it is.

```
Private Sub Winsock1_DataArrival(ByVal bytesTotal As Long) ' this is where all winsock data arrives
  Dim returndata As String    ' these are variables associted with the data received from perl server and the manipulation of that data
  Dim infostr1 As String
  Dim SplitInfo() As String   ' Returned data text manipulation string
  Dim SplitInfo1() As String  ' Returned data text manipulation string
  Dim intX As Integer
  Dim intY As Integer
  Dim intZ As Integer
  Dim statusval As Long
  Dim testb As String
  Dim testa As String
  Dim statusbinary As String
  Dim Splitinfoline() As String
  Dim infoofinterest As String
  Dim board4a As Long
  Dim imageset As Integer 'Our counter.
  Dim boardset As String
  Winsock1.GetData returndata, vbString ' get any data sent by Perl server
```

VB6 Code adding the functionality

```
  DataIn.Text = "Data from Perl Server " & returndata  ' debug show returned data
  testa = Mid$(returndata, 1, 1)  'Returns first character
 If InStr("0123456789", testa) Then  ' test to see if its a status or info message
    SplitInfo1 = Split(returndata, ",")  'its a status message split each value
    board0binary = DecimalToBinary(SplitInfo1(0))  ' convert Decimal integer  to
binary 8 bit word
    board1binary = DecimalToBinary(SplitInfo1(1))  ' convert Decimal integer to
binary 8 bit word
    board2binary = DecimalToBinary(SplitInfo1(2))  ' convert Decimal integer to
binary 8 bit word
    board3binary = DecimalToBinary(SplitInfo1(3))  ' convert Decimal integer to
binary 8 bit word
    Binary0.Text = board0binary  '  debug show binary value for each board
    Binary1.Text = board1binary  '  debug show binary value for each board
    Binary2.Text = board2binary  '  debug show binary value for each board
    Binary3.Text = board3binary  '  debug show binary value for each board
    boardset = board3binary & board2binary & board1binary & board0binary
    imageset = 1  ' set counter
    Do While imageset <= 32
        led(imageset).Visible = Mid$(boardset, 33 - imageset, 1)   ' turn on
channel leds board 1
        imageset = imageset + 1
    Loop
    Else
    SplitInfo = Split(returndata, "#")  ' splits the info string from perl server
    For intX = 0 To UBound(SplitInfo)
      testb = Mid$(SplitInfo(intX), 1, 1)  'Returns prefix character
        If testb = "B" Then   ' if "B" then its a board command else ignore
           Splitinfoline = Split(SplitInfo(intX), "|")
           infoofinterest = Splitinfoline(5)
            If Splitinfoline(1) = 0 Then
              B0Channel.AddItem infoofinterest   ' build the listbox for board 0
from our info data
            ElseIf Splitinfoline(1) = 1 Then
              B1Channel.AddItem infoofinterest   ' build the listbox for board 1
from our info data
            ElseIf Splitinfoline(1) = 2 Then
              B2Channel.AddItem infoofinterest   ' build the listbox for board 2
from our info data
            ElseIf Splitinfoline(1) = 3 Then
              B3Channel.AddItem infoofinterest   ' build the listbox for board 3
```

Chapter 10

```
from our info data
        End If
      End If
 Next
End If
Timer4.Enabled = True  ' wait a tad for data to settle
 End Sub
```

First we test to see if the initial characters returned are numeric, as we would expect in a status message. If it is a status message we then change it from decimal to binary, so that it equates with our board channels. In the binary just 8 bits are shown with a 1 indicating that the channel is on and a 0 indicating that a channel is off.

Once the binary has been decoded, it is used to set the led images so to show channels that are in the 'on' state as lit led's.

If the message is an info message then it is # delimited , so we strip away the hashes and split the data into phrases displaying those in the the board info lists, for the respective boards 0 to 3.

Now the last section of Code consists of two functions.

- Private Function ChangePBFForeColour is a function we call to change the colour of the vertical VU meter.
- Function Public Function DecimalToBinary is exactly that. We give it a decimal value as an argument and it returns the Binary equivilent value as a string.

```
Private Function ChangePBForeColour(ByVal hwnd As Long, ByVal lColor As Long)
    'Function to Change colour of VU meter bar
    SendMessage hwnd, PBM_SETBARCOLOR, 0, ByVal lColor
End Function
Public Function DecimalToBinary(nmr As Variant) As String
 ' Function to convert decimal value to binary byte string
Do Until nmr <= 1
    DecimalToBinary = Trim(Str(nmr Mod 2)) & DecimalToBinary
    nmr = Int(nmr / 2)
Loop
    If nmr = 1 Then DecimalToBinary = "1" & DecimalToBinary
        DecimalToBinary = Right("00000000" & DecimalToBinary, 8)
End Function
```

VB6 Code adding the functionality

That was quite a marathon, right?

It is one thing writing code, because you're thinking all the time, and time then goes fast. But, when you're copying code it's a real pain…, but I guess many of you did a cut and paste from the www.elektor.com support page. If you did, then do at least read the coding to see the comments, they will give you an idea of what's taking place.

As it is good practice, I urge you to copy the entire folder where you have saved your project to and put a copy on a usb stick or another PC, just as a backup. You don't want to lose all that hard work!

> **TIP!**
> When training SAPI, read the training text at normal speed.
> Do not emphasize any word, or attempt to speak in a clearer than normal voice or tone. In order for the system to recognize your voice you must train it using your normal, natural voice and tone.

Time for us to test the program.

To test you must have installed the Perl side of things including the dch-serv3.pl. If you have not you must do this now else it isn't going to happen!

First we must ensure that the xml file is saved to the default location "c:\rh2.xml". It should be as it would have been created by the Perl server program in real time, and the batch file c:\launchdch-serv.bat. must also be in the correct place. This is really important! The program will not function and may even crash on loading if it doesn't find this file and the file content is not correct.

Make sure your microphone is plugged in and that it has been calibrated, (if you haven't previously calibrated it, go to the control panel, Speech application and check your sound levels). Whilst there you may take the opportunity to do five minutes worth of training, it makes all the difference, Your PC stores a profile of you by learning your voice nuances which DRAMATICALLY improves accurate recognition. I did about 4 hours worth of training and can state it's really worth it! *'Nuff said'*.

Click on the Run menu on the IDE and select Start with full compile. Why? Well, because you are human and you will have made typo's, loads of them, a hyphen here a comma there, a variable, miss-spelt, a capital where it should be lower case.

You get the picture?

Chapter 10

Assuming ……………….. you now don't have any typo's and all the components are on the form, the program will compile without complaint and run. So up comes the GUI with the K8055 graphic and you will notice that the power led is on, but there's not much else going on that's visible.

Let's make things happen. Click on Audio level checkbox so it's unchecked, then click it again to check it. You will see a constant stream of audio levels in the Recognised Phrase and Events Frame.

The system is working. Rachel is alive!

Best try her out. Say "Rachel weather" into the microphone.

The entire Recognised Phrase and Events Frame will turn green indicating a recognised Phrase and the Caption of that frame will say "Rachel weather", which is the recognised phrase.

In the Value sent box you will see a message stating that there is no socket connection, No, she's not wrong, we just haven't yet got around to writing the Perl socket server that sets up the socket.

That concludes the VB6 part, which is the hardest part to get your head around.

Chapter 11

The Perl Server

Welcome to the hub of our wheel. On the periphery there may well be fancy-Dan speech recognition enabled VB scripts with all the bells, whistles and whizzy graphical user interfaces – but somewhere the heavy lifting needs doing … (is it really heavy? Not really, and we will walk through the relevant code, shortly).

As with most projects, things could be done in a variety of ways. Could VB have been used to talk directly to the USB-connected electronics? Of course it could, via the provided dll files.
Could Perl have been used to perform the speech recognition and graphical user interface? Of course it could, via the Win32::SAPI5 and Win32::GUI (or Tk) Perl modules. Compared to doing all the work in one program, the way we have chosen to do it, however, is more interesting (you get to play with more toys!) and more flexible (The Perl server, for example, is 'client-agnostic' – it doesn't care if you have a VB front-end, or a webpage or another Perl script or an iPhone talking to it).

At a high level, the server's functions include:
- Read a plain-text configuration file
- Build the XML grammar file for the VB Rachel program
- Read the device status file
- Connect to the K8055 USB board and set initial values
- Listen on a TCP/IP socket for 'Rachel calling'
- Send instructions to the K8055 USB board (to switch a light)
- Update the device status file
- Speak command confirmation out loud

Keep Rachel appraised

I can already hear the calls for 'Whoa! Slow down. Don't you realise I've never used Perl before? I can't do that!'
Well, perhaps not if you were on your own. Luckily, you're not. So read on and we'll explain as we go.

Assuming that you installed Perl earlier and configured the additional repositories we now need to execute the Perl Package Manager (PPM) again to install or check a few more modules. After a successful install of Perl, the path to the Perl executables (the 'bin' folder) will be on our system path. Open a windows command prompt and type: 'ppm' followed by enter. Hopefully this will launch the

Chapter 11

GUI of PPM. If it doesn't, then my first instinct would tell me that the windows command prompt that you have open right now was already running before you installed Perl. Am I right? Nothing psychic here – just experience. The windows command prompt inherits the system path at the time it is launched – and it doesn't get updated. Try simply closing the windows command prompt and then opening a new one.

Once PPM is running we need to ensure that we have the various Perl modules installed that we require for this server program. Ensure that the 'View all Packages' button has been pressed at the top left of the PPM window. This will show a list of all packages available in all of the configured repositories, (whether they are installed on your machine or not). Packages not installed have a grey icon to their left and installed packages have a yellow one.

Executing a Perl script will soon tell you if you have the necessary packages installed, but looking at the first few lines of the listing will normally tell you what you need to know. So, with that in mind let's look at the 4 'use' lines which immediately follow the 'shebang' and 'use strict' lines that we discussed earlier.

Here we go; lets look at the script in detail.

0. `#!c:/perl/bin/perl.exe` This line is sort of optional, except on some Unix machines, but we add it for completeness. It just tells the operating system where to find the perl compiler.

1. `use strict;` It is a Perl pragma which essentially keeps you from shooting yourself in the foot. It tells Perl to force you to declare your variables, use Subs and refs correctly, and basically make you write your programs correctly. In short, if it isn't written in the correct syntax it won't compile and will winge about every little thing it finds. However, if you don't use it, debugging code becomes a nightmare and your programming gets sloppy!!

2. `use IO::Socket::INET;` provides an object interface to creating and using sockets

3. `use Win32::API;` With this module you can import and call arbitrary functions from Win32's Dynamic Link Libraries (DLL).

4. `use Win32::OLE;` This Perl module accesses Microsoft's OLE (Object Linking and Embedding, pronounced "olay") it lets one program use objects from another program.

When an application makes an OLE object available for other programs to use, that is called OLE automation. The program using the object is called the controller, and the program providing the object is called the server. OLE automation is guided by the OLE Component Object Model (COM) which sets out and specifies how those objects must behave if they are to be used by other programs.

The Perl Server

There are two different types of OLE automation servers:
- In-process servers are implemented as dynamic link libraries (DLLs) such as the K8055 DLL and run in the same process space as the controller.
- Out-of-process servers are different, they are standalone executables that exist as separate processes. The Win32::OLE module lets your Perl program act as an OLE controller. It enables Perl to be used in place of other languages like Visual Basic or Java to control OLE objects. This module allows all OLE automation servers to be available as pseudo Perl modules.

You can think of it as a wrapper controlled by Perl that exposes all the functions of the OLE Object.

In this script we use win32::OLE to access and control the Microsoft SAPI(Speech Recognition) program features.

5. use Win32::Process; Perl extension to get all processes and their PID on a Win32 system.

This is where the main action takes place. We let VB6 do the donkey work of voice recognition and a little graphical interface, but this perl server script is what makes it all happen.

How, what, when?

Our VB6 program carries out the recognition and then if successful it sends a string via Winsock* to our dchserv3.pl script.

We have listed the script in its entirety, and once again we've added line numbers. Remember ,we have placed line numbers in front of the code shown above, in order that we can refer to the specific lines during our explanation of the code, however do not add line numbers in front of your code, it will not work!

```
use IO::Socket::INET;
use Win32::API;
use Win32::OLE;
use Win32::Process;
```

These are the packages whose statuses we need to verify – so, one at a time we type their name into the search box at the top of PPM. The list of packages will shorten as we type until eventually we end up with the item we want or (hopefully not) an empty list. At this point we should point out that while searching for packages in PPM it is normal to replace the :: with a dash so that, for example, Win32::API becomes Win32-API. Both, however, do work identically in the search box. From the

Chapter 11

results list you have in front of you now it is possible that PPM is offering you more than one version of each package. In the 'available' column it will be showing you the version number – the general rule here would be the higher the better. If the version numbers are the same, then it simply means that the requested package is available from more than one of your configured PPM repositories.

Highlight the item(s) you are interested in and then press the button to the right of the PPM search box, 'Mark for install'. PPM will remember the 'marked' item while you zoom off to check for other packages you may need. When you are finished you can press the green right-facing arrow, 'Run Marked actions' and the download and install procedure will commence. Some packages have prerequisites; PPM will be aware of these and it will offer to download and install those for you also.

Once this is done the Rachel server program should be ready to run, but just before you do, we need to configure some networking bits n bobs. Firstly, in the server script you will find the TCP/IP port number (5000) on line 95 and the LocalHost IP address on line 97.

The port number must match the one that Rachel is using in the VB code – they both default to 5000, so update in both places if you need to. The IP address can be set to the localhost address, 127.0.0.1, but this would mean that the server only 'listens' on the localhost – what's wrong with that? Actually, Nothing, as long as your client is on the same machine. If you want your server to be contactable from another machine then the IP address used must be that of an actual network card.

Typically PC's only have a single card installed and usually their address is allocated via DHCP. I'll assume that you know how to ascertain the IP address (hint: ipconfig) and you can enter it accordingly into the Perl server script. If you prefer, then the machine's name may be entered rather than the IP address, and as long as this name is 'ping-able' then the socket will setup correctly.

How do we install the stuff like stub,, status.dat, config? Now you will recall we use a config file (Chapter 9). Well, that has to be in its proper location - so if you haven't done so write it out or cut and paste from the support page and place it in the perl directory, whilst we are on that subject there are a couple of other files that need to be created and placed in the same directory. Namely, stub and status.dat, examples of both can be found in appendix E and F, respectively

Now you can run it, if you wish. Nothing will happen mind you, unless you send commands via Rachel or another mechanism. But running it will at least confirm to you that your Perl configuration is good to go. Open a windows command prompt and navigate using the 'cd' command to the directory where your copy of the Perl server lives and then type:

perl dchserv3.pl

followed by enter.

The Perl Server

If all your kit is good then you will see a series of confirmation messages containing a timestamp, port number confirmation, initial board settings and a 'Listening for client…' message. If something has gone wrong then the chances are that the error message is reasonably helpful and will point you towards the offending item, usually a missing PPM module.

If it's running and now is not the time for you to learn how or why, then skip to the next chapter. Otherwise, hang around for the deep dive.

The first thing that the server does is create a grammar file for Rachel to work with.

> Why does the server create this grammar file for Rachel? Can't she do it herself? Glad you asked! Well, it's not because she's lazy!
>
> The whole system is driven by a master configuration file, 'rachel-server.config', and it's a relatively simple human-readable, notepad-editable file. The individual fields are delimited by the pipe char (ASCII character 124). A few sample lines:
>
> B|2|2|Light|Office|Office one
>
> B|2|1|Light|Games Room|Entrance
>
> B|2|0|Switch|Garden|Gates
>
> S|9001|weather
>
> We need the Perl server to parse this file to get vital bits of information, for example, which K8055 board (you may plug 4 into each PC) and which channel (each K8055 has 8 output channels) any given light is attached to. Rachel doesn't need to know all that, nor does she have the ability to read this file unless we are prepared to code the same file parser twice – once in Perl and once in VB. We're not.
>
> So, the Perl server reads and captures all of the configuration data. It will then pass only the essential information to Rachel when she requests it.

The grammar file is built by the server using a very primitive 'template' system. What's that then? Essentially, it means that we load the static parts of the finished file from a pre-created 'stub' file, and then we drop the dynamic content right into it before we write it out for Rachel's usage.

So, what's this stub file then? In our case, it looks a lot like this: (Full listing in appendix F)

Chapter 11

```
<GRAMMAR LANGID="409">
  <DEFINE>
#DEFINES#
    <ID NAME="CmdType" VAL="0001"/>
    <ID NAME="Commands" VAL="0002"/>
  </DEFINE>
  <RULE ID="Commands" TOPLEVEL="ACTIVE">
    <P>#COMMAND_WORD#</P>
    <RULEREF REFID="CmdType" />
  </RULE>
  <RULE ID="CmdType" >
    <L PROPID="CmdType">
#PHRASES#
    </L>
  </RULE>
</GRAMMAR>
```

The content is not really relevant to us right now, but suffice to say that this is the very precise syntax required by the recognition engine. Perhaps you have noticed the 3 placeholders which will receive our dynamically created content?

```
#DEFINES#
#COMMAND_WORD#
#PHRASES#
```

Both #DEFINES# and #PHRASES# will be replaced by multiple lines of content, whereas the #COMMAND_WORD# will normally be replaced by a single word- the "get the computer's attention" word during speech recognition – in our case the word is "Rachel".

"Rachel, Office Light One On"

For each light or device we are controlling we only have one line entry in our master configuration file:

B|2|2|Light|Office|Office one

Rachel, however, will need 2 commands to be able to operate this device – an ON and an OFF. The Perl server is therefore also responsible for creating the multiple entries in the grammar file.
Once the grammar file is completed it will look more like:
(Where the ellipsis (…) represents other content removed for brevity)

```
<GRAMMAR LANGID="409">
  <DEFINE>
    ...
    <ID NAME="DEF1221" VAL="1221" /> <!-- Office one +on -->
    <ID NAME="DEF1220" VAL="1220" /> <!-- Office one +off -->
    ...
    <ID NAME="CmdType" VAL="0001"/>
    <ID NAME="Commands" VAL="0002"/>
  </DEFINE>
  <RULE ID="Commands" TOPLEVEL="ACTIVE">
    <P>+Rachel</P>
 <RULEREF REFID="CmdType" />
  </RULE>
  <RULE ID="CmdType" >
    <L PROPID="CmdType">
       ...
      <P VAL="DEF1221">Office one +on</P>
      <P VAL="DEF1220">Office one +off</P>
       ...
    </L>
  </RULE>
</GRAMMAR>
```

So, let's take a look at the code that's got us to this stage.

`my $stub="";`	Define a variable named $stub and assign it an empty string. All scalar variables (ones which can only hold one value at a time) start with the $ sign. Variables need to be declared within the scope of their usage if the "use strict" pragma is in effect. It is a good habit. Since this declaration is not inside nested code nor is it within a subroutine, the use of the 'my' makes the variable global. It is available for use throughout the entire program.

Chapter 11

open(STUB,"<stub01.xml")||die "cannot open xml stub";	Open a file named "stub01.xml" for input. The < symbol signifies input, a > symbol is output. For the purposes of performing actions on this file within the script give it a filehandle (name) of STUB. If you can't manage this without problems (the double pipe means OR), die (exit the script) with a non-zero return code and a relevant message.
while(<STUB>)	<STUB> means read a record from the input file with filehandle STUB. Putting a while condition around it means that we will perform the actions found within the following curly brackets for each record read, until the end of file (EOF).
{ $stub.=$_; }	These are the curly brackets. They could contain 1 program instruction or many. When the file read function is performed the default place for the data to be stored is a special variable named $_. We take the current value of the $stub variable and concatenate (join) it with the value stored in $_. So each time round the while loop, the value in $stub becomes longer and longer. Variables are quite happy to contain multiple lines of data and the line breaks are preserved by use of special escape characters.
close STUB;	When the while loop completes it means that we have read all the records in the file and we can close it.

`my @defines=();`	Define an array variable named @defines. All arrays (a data structure that lets you store multiple values in a single variable) start with the @ sign. Setting the value of the array to be empty parentheses means clear all the entries. Array entries are referenced using an index. $defines[0] would reference the first entry of the array (the index value starts at zero). When we are retrieving from an array we use the scalar $ rather than the array @ because what we are retrieving is a single scalar value.
`my @phrases=();`	Define an empty array
`my %cfg_val=();`	Define an associative array variable named %cfg_val. All associative arrays (also known as hashs) start with the % sign. Setting the value of the associative array to be empty parentheses means clear all the entries. Hash entries are referenced using a lookup key. This lookup key can be a number or a string. $cfg_val{"mickey"} would reference the entry whose lookup key was "mickey". The result could be a string ("mouse") or another data type. When we are retrieving from a hash we use the scalar $ rather than the hash % because what we are retrieving is a single scalar value.
`my @config=();`	Define an empty array
`my %lights=();`	Define an empty associative array
`my %office=();`	Define an empty associative array
`open(CFG,"<rachel-server.config")||die"cannot open rachel config";`	Open the master config file for input. Use a file handle of CFG. Exit with an appropriate message if it doesn't work out
`while(<CFG>)`	For each record in the config file

Chapter 11

{	Do the contents of the curly brackets
chomp;	Chomp – what a great word! Its function is to remove the carriage-return/line-feed combination from the end of a data item. Without any parameters like this, it will act upon $_. To chomp a variable other than $_, we simply put the variable name in parentheses, e.g. chomp($test). If the data item does not end with a CR/LF combo, then it will not be altered.
if(/^#/) { next; }	If the default variable ($_) starts with a # sign, then treat that line as a comment – do not continue further processing with this line and skip to the next. How does this line work? We use an 'if' condition to check a regular expression, and if it is true we use the 'next' command to skip to the next loop of the 'while <CFG>' above. Regular expression – now what's that? It's a topic that you could write a whole book about. Our regular expression starts and ends with a forward slash. These characters don't do anything, they are simply delimiters. So, that leaves our regular expression actual content as ^#. The caret ^, simply means 'starts with' and the # sign is the character we are looking for. So – it simply says, 'starts with #'.
push(@config,$_);	Add the contents of special variable $_ to the end of the array @config as a new entry

`my @fields=split(/|/);`	The split command acts on the default variable $_ unless you tell it otherwise. It uses a regular expression to locate the delimiters by which is should split. In the case we are splitting by the pipe character	. The pipe characters has a special meaning inside a regular expression (it means OR) but we actually want that character, so we have to 'escape' it with a back slash. So, for each piece of $_ separated by a |, add a field to the array @fields.
`if($fields[3] eq "Light")`	If the 4th entry in the @fields array is equal to the string "Light" (remember that the array index starts at 0 and that when retrieving a single data item (scalar) from an @ array, it is returned as a $ scalar)
`{ $lights{$fields[1].$fields[2]}=1; }`	B|2|2|Light|Office|Office one

Is an example of what $_ contains. The fields array will therefore contain 'B' at index 0, '2' at index 1, '2' at index 2, 'Light' at index 3, 'Office' at index 4 and 'Office one' at index 5.

This line of code is setting a hash value - you can tell because the scalar is referenced with curly brackets. The lookup for the hash is being set to $fields[1] concatenated with $fields[2] (that's what the full-stop does between the two variables).

In this example then, we are creating a lookup of "22", inserting it into the %lights hash with a value of 1. |
| `if($fields[4] eq "Office")` | If the 5th entry in the @fields array is equal to the string "Office" |
| `{ $office{$fields[1].$fields[2]}=1; }` | Same as '%lights' version above |

Chapter 11

if($fields[0] eq "C")	If the 1st entry in the @fields array is a "C" (if it is a 'C'ommand word)
{	Assume that the <CFG> line contained: C|COMMAND_WORD|+Rachel
$stub=~s/#$fields[1]#/$fields[2]/;	Use a regular expression to replace content in $stub. We are replacing #COMMAND_WORD# with +Rachel.
next;	There is nothing else to do for this CFG record type, so skip to the next one
}	End of condition processing
my $val="1$fields[1]$fields[2]";	Make a new variable $val, and populate it with the concatenation of "1", and the 2nd and 3rd entries of @fields
if($fields[0] eq "B"	If the 1st entry in the @fields array is a "B" (if it is a 'B'OARD command)

{	Assume that the <CFG> line contained: B|2|2|Light|Office|Office one
my $state=1; # ON	Create a new variable $state and give it a default value of 1
my $state_word="on";	Create a new variable $state_word and give it a default value of "on"
if($fields[3] eq "Switch") { $state_word="open";	If the 4th entry of @fields is equal to "Switch", then change the value of $state_word to be "open" rather than "on". Lights are on or off, switches are open or closed
}	End of condition

The Perl Server

`add_grammar_item($val,$state,$fields[5],$state_word);`	Call the subroutine 'add_grammar_item' and pass to it the 4 variables inside the parentheses
`$state=0; # OFF`	Set value
`$state_word="off";`	Set value
`if($fields[3] eq "Switch") { $state_word="close";`	If the 4th entry of @fields is equal to "Switch", then change the value of $state_word to be "close" rather than "off".
`}`	End of condition
`add_grammar_item($val,$state,$fields[5],$state_word);`	Call the subroutine 'add_grammar_item' and pass to it the 4 variables inside the parentheses
`next;`	Skip to next <CFG> line
`}`	End of condition
`if($fields[0] eq "S")`	If it is a 'S'cript command
`{`	
`add_grammar_item($fields[1],"",$fields[2],"");`	Call the subroutine 'add_grammar_item' and pass to it the 4 variables inside the parentheses
`next;`	
`}`	
`if($fields[0] eq "M")`	If it is a 'M'ulti/'M'acro command
`{`	
`my $state=1;`	
`my $state_word="on";`	

Chapter 11

add_grammar_item($fields[1] ,$state,$fields[2],$state_ word);	Call the subroutine 'add_grammar_item' and pass to it the 4 variables inside the parentheses
$state=0;	
$state_word="off";	
add_grammar_item($fields[1] ,$state,$fields[2],$state_ word);	Call the subroutine 'add_grammar_item' and pass to it the 4 variables inside the parentheses
next;	Skip to the next loop iteration
}	
}	
close CFG;	Close the master configuration file. To do this we use the filehandle.
my $defines=join("",@defines);	Join is the opposite of the split command. It takes all the entries in the @defines array and joins them back together into a single scalar string $defines. The first parameter of the join instruction is the delimiter that you wish to be put between the values. This could be a space, a comma, a new line – in fact anything at all – or, as in our case, nothing!
$stub=~s/#DEFINES#\ n/$defines/;	Use a regular expression to replace content in $stub. We are replacing #DEFINES# with the contents of the $defines variable (this is our simple 'templating' system in action!)
my $phrases=join("",@ phrases);	Join the @phrases array in the same way as we did the @defines

$stub=~s/#PHRASES#\n/$phrases/;	Use a regular expression to replace content in $stub. We are replacing #PHRASES# with the contents of the $phrases variable
open(GRAMMAR,">c:/rh2.xml")||die"cant open new grammar";	Open a file for output (the > sign tells us this). Give it a filehandle of GRAMMAR. Exit with a message if it doesn't work out.
print GRAMMAR $stub;	Print the contents of the $stub variable to the GRAMMAR filehandle we have just opened
close GRAMMAR;	Close the file
my @status=();	Create a new empty array
open(STATUS,"<rachel-server.status")||die"cannot open rachel status";	Open STATUS file for input
while(<STATUS>)	For each record…
{	
chomp;	☺ There's that word again! Get rid of those pesky carriage return and line-feeds!
@status=split(/,/);	Split $_ into @status using a comma as the delimiter
}	
close STATUS;	

Chapter 11

`my $OpenDevice=Win32::API->new('K8055D','OpenDevice','I','I');`	Create a new object called $OpenDevice. This object is for calling a function named "OpenDevice" which exists within a dll file provided by the K8055 manufacturer. Perl cannot call dll functions directly, so we are making use of the Win32::API package that we installed earlier. The 'new' method accepts 4 parameters within parentheses. They are: 1. The name of the dll without the extension. The dll must be in the same directory as your Perl script or it can be located somewhere of the windows PATH. 2. The name of the function within the dll 3. The input parameter list specifies how many arguments the function wants, and their types. In this case the 'I' represents one integer input value. 4. The type of the value returned by the function. In this case the 'I' represents one integer return value. The function names and input/return types can normally be ascertained from the C language header files (.h) that are provided by the dll author (if they want you to be able to control the dll in your own code!)
`my $CloseDevice=Win32::API->new('K8055D','CloseDevice','','');`	Similar to OpenDevice but no parameters are required to the 'CloseDevice' function call itself.
`my $WriteAllDigital=Win32::API->new('K8055D','WriteAllDigital','I','I');`	Similar to OpenDevice

The Perl Server

my $port=5000;	Create a new variable $port and give it the value 5000
my $socket = new IO::Socket::INET (LocalHost => '192.168.0.03',LocalPort => $port,Proto => 'tcp',Listen => 5,Reuse => 1)||die "Unable to create socket on port $port : $!\n";	Make use of the IO::Socket::INET package that we installed earlier to create a socket. Create a $socket object to point at the instance of the socket we are creating. Parameters required are: LocalHost - discussed earlier. LocalPort - set by $port above Proto - set to 'tcp' (the package supports other prototypes, 'udp' for example) Listen - gives the queue size for the number of client requests that can wait for an accept at any one time Reuse - Given a non-zero number, this option allows the local bind address to be reused should the socket need to be reopened after an error. If not, you may have to wait an unspecified amount of time for the port to become available (timeout) after an error, before you can restart your Perl server.
log_it("\n\nR&D Harwood 'RACHEL' Automation Server Initiated on port $port\n");	Call a subroutine for logging what's going on. It accepts a single parameter within parentheses
for (my $i=0;$i<4;$i++)	Create a loop which does something 4 times. a 'for' loop accepts 3 parameters: my $i=0 - setup a local variable (can only be used within the scope of this 'for' loop) $i<4 - continue looping while the value of $i is less than 4 $i++ - increment (add 1 to) $i

Chapter 11

`{`	
`WriteAllDigital($i,$status[$i]);`	Call a subroutine 'WriteAllDigital' and pass it 2 parameters (the value of $i, which is the loop counter and the value of the @status array entry which is indexed by the value of $i. This subroutine will actually communicate with the K8055 board and set its values.
`log_it("initial startup, setting board $i to $status[$i]");`	Log another message
`}`	
`my $HellFrozenOver = 0;`	Create a new variable which will monitor our infinite loop
`log_it("Listening for client....");`	
`my $client_socket = $socket->accept();`	Create the $client_socket object. This will be used to communicate with the client that was trying to connect.
`my $peer_address = $client_socket->peerhost();`	Get the client's IP address
`log_it("ah, there it is! $peer_address");`	
`while(!$HellFrozenOver)`	Loop while hell is not frozen over. Hell frozen over would be a value of 1. The exclamation mark makes this a negative test condition
`{`	
`my $data="";`	
`$client_socket->recv($data,1024);`	Receive up to 1024 bytes from the socket and store it in $data

The Perl Server

`chomp($data);`	☺
`if ($data eq "7777") {` `$HellFrozenOver=1; last;`	This is our bail-out of the infinite loop opportunity. If the client sends "7777" then jump out of the loop, close the socket and exit the Perl server
`}`	
`log_it("received: '$data'");`	
`my $response=&process($data);`	Pass the $data variable to the sub routine named 'process' and store the return value in $response

`$client_socket-`
`>send($response."\`
`n")||die"cant send back response";`
	Send the response, with a concatenated line-feed, to the client via the socket. Check if anything goes wrong.

`log_it("response sent - '$response'");`	
`log_it("waiting for next command");`	
`}`	
`$socket->close();`	Close the socket
`exit 0;# main process finished, now some subroutines`	Exit the script with a zero return code. Zero typically means that all went well and this return value can be checked by any calling program.
`sub process`	Start the sub routine named 'process'
`{`	

Chapter 11

`my $instruction=shift;`	Create a variable $instruction. Set it to a value provided by the 'shift' command. Shift command? If parameters are passed to a subroutine, they are sent as an array. It is the default array and it is named @_, similar to the default scalar $_. The shift command acts upon arrays. If it receives a parameter, then it acts upon that named array, else it defaults to @_. So, what does shift actually do? Well, it removes the first entry from the array and shuffles all the other values to the left (as it were). The value it removed is the return value and therefore in our example, $instruction will be set to that value.
`my $time=localtime;`	'Localtime' is a built in Perl function to return the time as a string. The format can be varied depending on the method of calling.
`if($instruction=~/^9/)`	Condition to check if the value of the $instruction variable starts with a 9
`{`	
`launch_ext($cfg_val{"$instruction"});`	Call the 'launch_ext' (launch an external) subroutine and pass it a value. In this case the value will be the result of a lookup into the %cfg_val hash. The key will be the value of $instruction. An example of a relevant <CFG> file line: S|9001|weather The lookup would be 9001 and the returned value will be 'weather', which will be passed to 'launch_ext'.
`return;`	Return from the subroutine

`}`	
`if($instruction=~/^8/)`	Condition to check if the value of the $instruction variable starts with an 8
`{   my $to_say="";`	$to_say is a new variable that we will use to get the Perl server to speak back to us. This will store the actual text that Rachel will say.
`if($instruction=~/^802/)`	Condition to check if the value of the $instruction variable starts with 802
`{    $to_say="Office lights ";`	Set the text
`while (my ($key, $value) = each(%office))`	For each entry in the %office hash get the lookup key and the value and store them in local variables $key and $value
`{`	
`action($key,substr($instruction,3,1));`	Call the subroutine 'action' and pass it 2 parameters. The first is the hash key and the second is the 4th character of the $instruction variable. How do we get the 4th character? We use the substr function (sub-string). 4th because it is an offset number and therefore starts at 0. The second parameter passed to the 'substr' function is simply the number of characters we want retrieved.
`}`	
`}`	
`if($instruction=~/^801/)`	Condition to check if the value of the $instruction variable starts with 801
`{    $to_say="All lights ";`	

Chapter 11

while (my ($key, $value) = each(%lights))	For each entry in the %lights hash get the lookup key and the value and store them in local variables $key and $value
{	
action($key,substr($instruction,3,1));	
}	
}	
if(substr($instruction,3,1) eq "0")	If the last character (4th) of the instruction is a 0 it means off, it is a 1 it means on
{ $to_say.="off";}	
Else	
{ $to_say.="on";}	
my $Process_Object;	Create an object to point to our process.

The Perl Server

`Win32::Process::Create($Process_Object,"C:\\Perl\\bin\\perl.exe","perl speak.pl \"$to_say\"", 0,DETACHED_PROCESS, "." ) || die "cant speak";`	Use the Win32::Process package to create a new process. Our main server script will launch the new process and will not wait for it to complete. If it is unsuccessful, exit with a message. The parameters to pass to the 'create' method are: $Process_Object as created above. String containing the full path to the executable we wish to call. The slashes must be windows style (ie, back slashes and they must be escaped, so each one becomes two). The command line of the program we are calling, including the parameters we are passing to it. So, in this case we are calling the perl executable to run a script called speak.pl – we are passing it a parameter of the text we want to speak. That text must be in quotation marks, and because our quotation marks are nested within other quotation marks they will need escaping with a back slash also. Inherit handle flag – 0 means no for us. Creation flags – we are a distinct detached process. Working directory of new process (the folder we are in)
`my $reply="";`	
`for (my $i=0; $i<4; $i++)`	
`{`	
`WriteAllDigital($i,$status[$i]);`	
`$reply.=$status[$i];`	

Chapter 11

`if($i<3) { $reply.=","; }`	Concatenate the reply with a comma unless it is the last one
`}`	
`write_status($reply);`	Call 'write_status' subroutine sending one parameter
`return $reply;`	Return from the subroutine passing a return value
`}`	
`if (exists $cfg_val{$instruction})`	Condition to see if an entry appears in the %cfg_val hash with the value of $instruction as the key.
`{`	
`my $to_say=$cfg_val{$instruction};`	If it does, get the value of it
`$to_say=~s/\+//g;`	Use regular expression to remove any + signs which are there (present in the grammar file to emphasise important words)
`my $Process_Object;`	
`Win32::Process::Create( $Process_Object,"C:\\Perl\\bin\\perl.exe","perl speak.pl \"$to_say\"", 0,DETACHED_PROCESS, "." ) || die "cant speak";`	Create a detached speaking process
`}`	
`my @parms=split(/,/,$instruction);`	Split the $instruction variable by comma and store the results in @parms array
`my $reply="";`	

The Perl Server

`if (uc($parms[0]) eq "STATUS") {`	Condition to see if the uc (upper case) value of the first entry in @parms is equal to the string "STATUS"
`$reply=&status(\@parms); } # pass array by reference`	Call the subroutine 'status' and pass a parameter by reference. Store the result in $reply. By reference means that we are 'sharing' the value passed with the subroutine rather than passing it a copy of it. So, if the sub updates the value, then the caller's copy is updated too.
`if (uc($parms[0]) eq "INFO") { $reply=&info(\@parms);`	
`} # pass array by reference`	
`if (uc($parms[0])=~/\d\d\d\d/)`	Condition to see if the upper case value of the first entry in @parms is 4 numeric digits. The regular expression is \d four times. The slash is to escape and therefore indicate special behavior of the 'd'. the 'd' in this context represents any single number 0-9
`{ if(exists $cfg_val{"$parms[0]"})`	
`{`	
`my $type=substr($parms[0],0,1);`	
`my $board=substr($parms[0],1,1);`	
`my $channel=substr($parms[0],2,1);`	
`my $state=substr($parms[0],3,1);`	

129

Chapter 11

`log_it("expands to: command type:$type, board num:$board, channel num:=$channel, bit state to:$state");`	
`my $temp=$status[$board];`	
`my $out = sprintf("%08d",unpack("B*", pack("N", $temp)));`	Set the $out variable to the formatted output of the sprint command. The sprint command can take many parameters; the first always describes the required output format. In our case "%08d" means pad a number to the left with zeros so that it is 8 digits long. The second parameter is that number to be padded which is calculated by the 'unpack' function. $temp contains a decimal number within the range 0-255 and we use the pack function to create a number of a known length. This known length is required for the unpack function which is being asked to convert that number into a binary string representing bits for us.
`if(substr($out,7-$channel,1) eq $state)`	Condition to see if the 'bit' we are interested in, is already in the state that we proposing switching it to. So, if the light is already on, don't do it again. Our bits are stored backwards, so we need to perform the 7-$channel calculation. Our $channel has values 0 to 7.
`{`	
`print "already in that state\n";`	
`}`	
`Else`	

`{`	
`my $a=2**$channel;`	Create new variable $a and set its value to 2 raised to the power of $channel
`if($state eq "1")`	
`{ $temp+=$a; } # detected ON request`	If it's an 'on' request, add the $a value to $temp which is storing the status of our board.
`Else`	
`{ $temp-=$a; } # detected OFF request`	If it's an 'off' request, subtract the $a value from $temp which is storing the status of our board.
`$status[$board]=$temp;`	Update the board status array entry with our new value
`WriteAllDigital($board,$status[$board]);`	And then call the 'WriteAllDigital' subroutine to actually update the board
`}`	
`$reply="";`	Reset the value of our reply variable before we calculate the new value
`for (my $i=0; $i<4; $i++)`	Loop through values 0 to 3
`{`	
`$reply.=$status[$i];`	Concatenate the status of each board onto our reply string
`if($i<3)`	
`{ $reply.=","; }`	Again, we don't want the comma added after the last entry
`}`	

Chapter 11

`write_status($reply);`	Call subroutine 'write_status' and send $reply as the parameter
`}`	
`Else`	
`{ print "Invalid command format, must be 4 digits - received '$parms[0]'\n";`	We didn't like what we received so print a message
`$reply="-1";`	And then set the reply to be -1. The client should check for this and take appropriate action – it means that the server didn't do anything.
`}`	
`}`	
`return $reply;`	Return from the subroutine passing $reply as a return value
`}`	
`sub log_it`	Start the 'log_it' subroutine
`{`	
`my $time=localtime;`	
`my $msg=shift;`	Set a variable $msg by using the shift function on the default array

The Perl Server

`$msg=~s/^(\n+)//;`	Remove any extra blank lines at the start of a message. The regular expression is ^ (starts with)
	\n+ any number of carriage-return/line-feeds and replace them with nothing. The parentheses around the \n+ capture the output before replacing it.
`my $blanks=$1;`	The captured data from the above regex (regular expression) is stored in a variable $blanks. We do this to prevent blank lines appearing between the timestamp and the message text itself
`print "$blanks$time - $msg\n";`	Print the blank lines (if any), the time and the message body
`}`	
`sub status`	Start the 'status' subroutine
`{`	
`my $parms=shift;`	Set a variable $parms by using the shift function on the default array.
`my $parm_count=@$parms; # get number of array elements`	Set a new variable $parm_count to be the number of entries in the @parms array which was passed by reference
`if ($parm_count > 1) { log_it("STATUS command received with unrequired parameters. ignoring them.");}`	Issue diagnostic message
`my $reply="";`	
`for (my $i=0;$i<4;$i++)`	
`{`	
`$reply.=$status[$i];`	

133

Chapter 11

`if($i<3){$reply.=",";}`	
`}`	
`return "$reply";`	
`}`	
`sub info`	Start subroutine 'info'
`{`	
`my $parms=shift;   # receive array reference`	
`my $parm_count=@$parms;   # get number of array elements`	
`if ($parm_count > 1) { log_it("INFO command received with unrequired parameters. ignoring them.");}`	
`my $return_info=join("#",@config);`	Build the '$return_info' value by joining the @config array entries with a # sign between each
`return $return_info;`	
`}`	
`sub OpenDevice`	Start subroutine 'OpenDevice'
`{`	
`my $device=shift;`	Receive the device number (board number) from the shift function

`my $rc=$OpenDevice->Call($device);`	Call the K8055 dll function 'OpenDevice' and send it the board number. This will start communications with the board itself via USB. The return code of that connection request is captured into $rc
`}`	
`sub CloseDevice`	Start subroutine 'CloseDevice'
`{`	
`my $rc=$CloseDevice->Call();`	Call the K8055 dll function 'CloseDevice'. It does not require any parameters as only one board may by connected at any given moment.
`}`	
`sub WriteAllDigital`	Start subroutine 'WriteAllDigital'
`{`	
`my $device=shift;`	
`my $value=shift;`	
`OpenDevice($device);`	Call our Perl subroutine to open the device and start communications with the K8055 board
`my $rc=$WriteAllDigital->Call($value);`	Call the K8055 dll function 'WriteAllDigital' and send it the new 8 bit number (0-255). This will set all 8 output channels on the currently connected board. The return code of that request is captured into $rc
`CloseDevice();`	Call our Perl subroutine to close the device and stop communications with the K8055 board
`}`	
`sub add_grammar_item`	Start 'add_grammar_item' subroutine
`{`	

Chapter 11

my $def=shift;	
my $state=shift;	
my $desc=shift;	
my $state_word=shift;	
$state_word=" +".$state_word unless $state_word eq "";	If the $state_word isn't empty set it to be a space and a plus sign concatenated with the current value of $state_word
if(exists $cfg_val{"$def$state"})	Check to see if the %cfg_val hash contains a value with the concatenation of $def and $state as its key
{	
print "warning....value 'defstate' already exists. ignoring dupe\n";	
return;	
}	
$cfg_val{"$def$state"}=$desc.$state_word;	Add an entry to the %cfg_val hash which has the concatenation of $def and $state as its key and the concatenation of $desc and $state_word as its value

The Perl Server

push(@defines, qq`\t\t<ID NAME="DEF$def$state" VAL="$def$state" /> <!-- $desc$state_word -->\n`);	Add a new entry to the @defines array. The entry value is made up from a series of literal text items and some variable values. Instead of starting the value with quotation marks we have used 'generalised quotes' which consists of 'qq' followed by a string contained within back ticks. These are an alternative to using quotation marks, especially useful if your string itself is actually going to contain quotation marks. We are building a line of XML here for our templating system.
$desc=~s/\&/\&/g;	The ampersand is a special character within regular expressions and needs escaping with a backslash. The regular expression itself is replacing all occurrences of the ampersand sign with the XML escape sequence for an ampersand which is "&". The regex has the 'g' modifier on the end meaning 'global' (i.e., do it more than once if necessary)
push(@phrases, qq`\t\t\t<P VAL="DEF$def$state">$desc$state_word</P>\n`);	Similar to the above but for the phrases portion of the XML template
}	
sub launch_ext	Start the 'launch_ext' subroutine
{	
my $name=shift;	Receive the name of the external process we want to launch. 'weather' for example
my $parm="";	Create a new variable $parm to hold any (optional) parameters

Chapter 11

`if($name=~/(.*?)=(.*)/) {` `$name=$1; $parm=$2; }`	Condition to see if the $name variable contains an equals sign. This is the delimiter we have used if the external command is to receive a parameter. A good example of this is the music command. music=play the word 'music' will be stored in $name and 'play' will be stored in $parm
`my $Process_Object;`	Create a process object (the same as we did for the speech process)

`Win32::Process::Create(`
`$Process_Object,"C:\\Perl\\`
`bin\\perl.exe","perl $name.`
`pl \"$parm\"", 0,DETACHED_`
`PROCESS, "." ) || die "cant`
`launch $name";`
	Create a detached process and don't wait for it to complete. This process will run the perl interpreter with the name of the script provided in $name and it will pass the paramaters within $parm to it.

`}`	
`sub action`	Start the 'action' subroutine
`{  my $item=shift;`	
`my $state=shift;`	
`my $board=substr($item,0,1);`	Get the $board number from the first character of $item
`my $channel=substr($item,1` `,1);`	Get the $channel number from the second character of $item
`my $temp=$status[$board];`	Get the status of board with number $board and store it in a temporary variable

`my $out = sprintf("%08d"`
`,unpack("B*", pack("N",`
`$temp)));`
	Pack and unpack the number into a binary string

`if(substr($out,7-$channel,1) ne $state)`	Use that binary string to see if the channel is already in the requested state
`{`	
`my $a=2**$channel;`	Create new variable $a and set its value to 2 raised to the power of $channel
`if($state eq "1")`	
`{ $temp+=$a; } # detected ON request`	If it's an 'on' request, add the $a value to $temp which is storing the status of our board.
`Else`	
`{ $temp-=$a; } # detected OFF request`	If it's an 'off' request, subtract the $a value from $temp which is storing the status of our board.
`$status[$board]=$temp;`	Update the board status array entry with our new value
`}`	
`}`	
`sub write_status`	Start the 'write_status' subroutine
`{`	
`my $reply=shift;`	Receive the status via the shift function
`open(STATUS,">rachel-server.status") || die"cannot open rachel status for write";`	Open the 'rachel-server.status' file for output
`print STATUS $reply,"\n";`	Print the value of $reply to the STATUS file
`close STATUS;`	Close the status file
`}`	

Chapter 11

So what does the Perl server program actually do?

Heck, it's more a case of what doesn't it do, let's try and go through it.
- The Perl server is first called by the VB6 Rachel program via a batch file.
- As the VB6 program starts it calls the batch file to start the perl server and waits for it to start running.
- The VB6 program cannot do anything until the perl server starts because it needs certain information from the perl server and it needs the grammar xml file that the perl server builds on the fly. So having got the start, command Perl starts the server script.
- The perl server starts it's tcp/ip socket process to allow communications with the outside world. Next, the perl server opens the config file, reads it and builds the grammar file, based upon the contents.
- The grammar file is then placed in the root of c: (its default location).
- Just as the grammar file has completed being built, the server program receives a request from the VB6 program for "info". This info request asks the perl program to look at the config file and extract the name part for each board and channel.
- This info is sent via a tcp/ip socket to the VB6 program and used to populate the four interface board text panes.
- The next request received from VB6 is a request for "status,..
  Status corresponds to the last known configuration of all the boards and their channels, whether on or off. This information is kept in a Status file which the perl server reads as it starts.
- The perl server sends the read status to VB6 via a tcp/ip socket and having completed its initial tasks sits back "listening" for the next command.
- The perl server receives its commands via a tcp/ip socket in strings of characters and digits up to five characters long".
- When a command string is received, the perl server reads the first digit to determine whether the command is for an interface board (B) ,a script command (S) or a macro command (M) .
- Based on what it finds here it branches to one of three possible avenues.

Avenue B

Having found a B command the perl program then looks at and decodes the remaining four digits of the string. Those four digits tell the perl program which interface board it is dealing with, what channel is selected and whether that channel should be switched on or off.

The perl server program then connects to the relevant interface board in question and sends it the command to turn on or off the relevant channel.

So what does the Perl server program actually do?

Once that channel has been changed, the perl server then tells Rachel to announce that she has carried out the task, then updates the status file to reflect the new state of the selected channel, finally sending a confirmation message back to the VB6 program via a tcp/ip socket that the task has been completed.

Avenue S

When a command S is encountered the perl server looks at the following string to determine the name of the script that is required to run and then sends off a command to perl to execute that script.

There is a bit of background wizardry going on here that I should mention, else Darren will get the hump.
When a command is sent to the perl server that calls for speech to be generated that speech command is placed in a queuing system. It would not be good if Rachel started speaking the weather and then received a command for the time and spoke over the top of herself, so all speech commands are queued.

Avenue M

The M command means macro, well sort of. What we mean here is that we want to send a command to an interface board or boards which tell a group of channels to either switch on or off. That may be, for example, all the ground floor lights or certain lights set up in a mood scenario.

The perl server reads the macro request, looks at the config file to get the address of each of the channels affected by the request and then produces a composite address so that the interface board or boards are all written to at the same time. The perl server then tells Rachel to announce that she has carried out the task, writes to the status file to update to the current state of each channel and finally reports back to Vb6 program that the command has completed.
Once any command has completed, the perl server goes back to listening on a tcp/ip socket for the next command.

Chapter 12

The K8055 (VM110)

For a computer to real world interface we are using the Velleman Usb Experimenters board.

There are dozens of other boards out there that do the same, but this is great if you want to build the electronics as well, because Velleman supply either in kit form or ready built. We will show you later on how to use another board, the PC-control Master/ slave setup which is just as good, and in certain situations even better. But we will describe the building of the K8055 system and include the full PC-control VB6 program on the Elektor support page.

I need to tell you a bit about his board, so that you can understand how we achieve what we do! What does it look like? There you go one k8055 board:

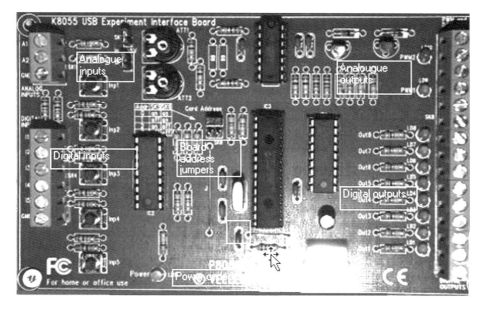

The K8055 interface board has 5 digital input channels (Left hand side connector block) and 8 digital output channels (right hand side connector block with leds 1-8 beside them). In addition, there are two analogue inputs (top left connector block) and two analogue outputs with 8 bit resolution (top right connector block with two leds).

Chapter 12

The number of inputs/outputs can be further expanded by connecting up to a maximum of four additional cards to the PC's USB connectors.
The lower led is a Power indicator led.
The function of the board is to take our commands from our program and turn on or off the appropriate channels. It takes our USB signals in/out via the USB port where they are decoded and fed through a Darlington Transistor Array and appear at the channel output as open collector outputs.

What's an Open collector output?

We tend to think of most outputs as a voltage, say +5v for on and 0v (zero) for off, but open collector works differently. The output essentially acts as either an open circuit (no connection to anything) or a connection to ground. So in order to use this we have to power our relay and use the output as a connection to ground, thus completing the circuit. Here's what I mean:

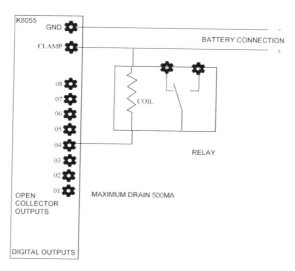

What's a Darlington Transistor Array?

It's an electronic chip that takes low level, low voltage computer signals (TTL) 0-5 Volts and turns them in to high voltage/ current capable signals. These higher voltage signals can be used to turn on and off relays and such, but on their own are not suitable for mains switching. We need a relay!
We do this by connecting a K8056 board to our K8055 and then we have capability of switching mains voltages and currents via 8 relays.
All communication routines are contained in a Dynamic Link Library (DLL) which we call from within our Perl script.

The K8055 (VM110)

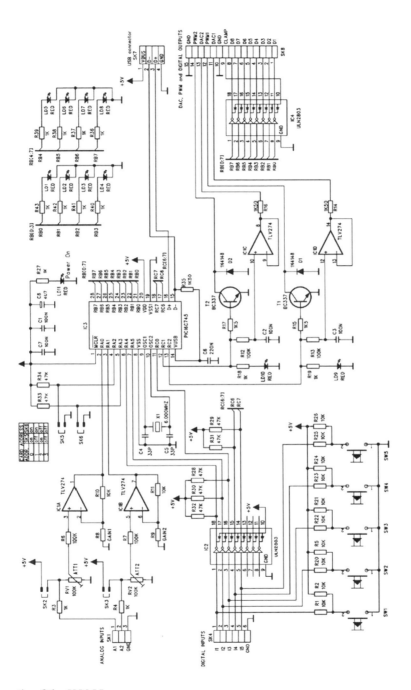

This a schematic of the K8055

The K8056

The K8056 is an eight channel relay card,

Specifications:
- 8 high quality relay contacts: 5A/230Vac max.(top right of board numbered 1 to 8)
- Relay outputs are transient suppressed using VDRs
- LED confirmation on each relay contact
- 8 drive inputs to use with open collectors or regular switches
- RS232 input to drive the card with computer or terminal: baud rate: 2400, no parity, 8 data bits, 1 stop bit, high or low impedance selection (10K or 1K)(bottom left of board two connectors marked RS232
- Power supply: 12Vac / 500mA (top left connector) It has its own onboard voltage rectifier and regulator.
- Dimensions: 160 x 107 x 30mm (6.3" x 4.2" x 1.2")

Things to note about this board.

We are not using the RS232 serial interface to control this board. We could easily, but so many PC and laptops these days don't have serial ports so it means buying an adapter. You might as well use the K8055 which has other features we can utilize, such as Analogue to Digital inputs and Pulse width modulation out, and easy usb connection.

Also next to the RS232 connectors there are two connectors for 12 volts out, we do not use this in our project. Ensure you don't connect to these by mistake when connecting to the k8056 inputs that are located alongside!

The switched loads are available from the top right connectors numbered 1-8, connected as live feed in switched feed out. The contacts available from this board are NO (Normally Open), and although the relay's on the board do have NC (normally closed) contacts they have not been bought to the connectors as available. You could easily modify the board and the usual mains voltage warnings apply; if in doubt, get it done professionally!

The K8056

> Caution: Although the Relay contacts are rated for mains voltages, there is no protection provided for the exposed live terminals and particularly on the underside of the board. See Chapter 3, Read and re-Read!!!If this board is to be used with mains voltages then care must be taken to ensure all exposed terminals are made safe by using a suitable enclosure to encase the entire board..The board should not be used "as supplied" with mains voltages. If you are not qualified, or are in any doubt, you should consult a qualified electrician.

If this board has mains voltages on it, it needs to be in an enclosure (Velleman supply a suitable one, example their catalogue number B8006) at all times when connected to mains voltages.

Nothing too magic about this board! 8 relays controlled by 8 opto-isolators, with 8 display leds. There is a little bit of wizardry with a RS232 input to drive the card with a computer or terminal

Chapter 13

PC-Control Master and Relay Slave

This is just one of the alternatives for producing computer to real world interfacing.

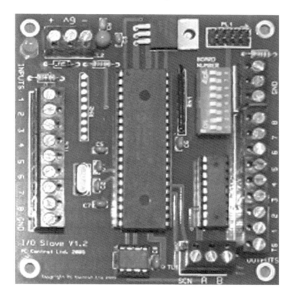

The PC_ControlOpto isolated Master Unit
This system is quite unique when compared with the Velleman K8055 and K8056 system, in so much as the actual relays can be placed anywhere on a two wire network, up to 1 kilometre away from the master. PC-Control Master uses a Master–Slave principle of operation.
In the system there is one Master Controller board which is connected to a computer via a standard USB lead and in turn can be connected to a maximum of 30 slave boards.
The job of the Master controller is to allow the PC to send and receive data to any of the slave boards.
The Master Controller has an on-board RS485 driver which provides communications with slave boards up to 1Km (0.62 Miles) away from the master controller.

The Master also has an on-board Microcontroller which is dedicated to performing two main functions:
1. Handling the USB interface and communications to the PC
2. Handling the RS485 serial communications to the slave board.

There are two different types of Master Controller available; the Non-isolated Master and the Isolated Master. The Isolated Master should always be used as it provides electrical isolation from the board and the PC, thus reducing the chances of 'frying' your PC. This isolation is actually a light emitting diode and a light sensor encapsulated in the same small 8-pin chip. This Opto isolator provides isolation for up to 2500 volts.

The Microcontroller and associated components on the Master are powered from USB whilst the RS485 communications components require a separate supply.

So we can think of the master as being a communications interface, sending commands to and from the slave modules, although we are not discussing it in detail in this book. The master and slaves are two way. That is, you can "Read" the slave inputs as well as "Send" to the slave outputs. This opens up another can of worms, but allows the user to read, say temperature sensors or contact switches and take action on what the input state is.

Each slave can be connected in parallel and/or can also be connected to other slaves in a star configuration. The only limitation is that no slave can be any more than 1Km from the Master.
This flexibility makes the distributed control system very suitable to a wide variety of installations.

PC-Control Master and Relay Slave

CONNECTING RELAY SLAVES TO PC-CONTROL ISO MASTER

Chapter 13

Relay Slave

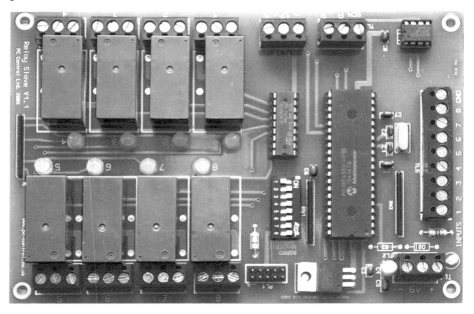

The Relay Slave has 8 digital inputs and 8 changeover relay outputs. With a master controller already connected to the PC this slave module can be up to a massive 1Km away connected only by a single pair of wires (Connections are made to the connector block top right marked scn A and B, the two wires required go A to A, B to B. If using shielded wiring connect the screening to scn)

Board Numbering: One last task is required before the Relay Slave can take part in the main control system and that is to allocate it a board number. It is necessary to allocate each board a unique Board Number so that commands and data from the Master Controller can be directed at the correct slave board. This is done by setting the blue DIL switches on the board labelled, funny enough "Board Number".

Relay Outputs: The 8 Relay outputs are available on 8 sets of 3-way screw terminals labeled simply 1 to 8. The centre terminal in each group of 3 is the common connection which is connected to the left hand terminal when the relay is not energized and the right terminal when energized. Left and right are determined when looking at the wire entry face of the terminals. Directly behind each relay is an LED indicator showing (when on) that the relay is energized. The contacts are rated for 6A at up to 250v AC (or 30v DC).

Caution: Although the Relay contacts are rated for mains voltages, there is no protection provided for the exposed live terminals and particularly on the underside of the board. See Chapter 3, Read and re Read!!!

If this board is to be used with mains voltages then care must be taken to ensure all exposed terminals are made safe by using a suitable enclosure to encase the entire board...

The board should not be used "as supplied" with mains voltages.

If you are not qualified or are in any doubt consult a qualified electrician.

Chapter 14

Perl Scripts for switching and talking

This section is just to show what can be done with Perl and SAPI.
With just a few lines of text we can make the computer speak anything, type this into any text editor and save as speak.pl

```
# This code is © R&DHarwood 2011. No part of this code may be copied, or used un-
less  a copy of the book PC Voice Control System by R&D Harwood is owned by that
user, and this notice is preserved in the coding.

#!c:/perl/bin/perl.exe
#speech starts here

use Win32::OLE;
my $v=Win32::OLE->new('SAPI.SpVoice');
my $voice="Rachel(British)SAPI5";
my @eee=$v->GetVoices();
$v->{voice} = $v->GetVoices("name = $voice")->Item(0) if $voice;

$v->Speak(" This is how easy it is ");
$v->Speak("For me to speak");
$v->Speak("Anything else I can do for you?");
exit 0;
```

If you have Perl loaded on your machine, just copy this file into your perl directory and from command prompt type perl speak.pl

Chapter 15

TTS Voices

TTS or **text-to-speech** is the digitized audio conversion of computer text into speech.
TTS software can "read" text from a program, document, Web page or e-Book, and then generate a synthesized voice through a computer's speakers/ sound system.

Windows operating systems have TTS programs built in to them, so we are going to hijack the TTS program and call it from ours. This will give us complete voice functionality.

Now, there are TTS voices and there are TTS voices, and it is for you to decide what you want to hear. Do you want a Robotic sounding voice or one that's near indistinguishable from human?
We wanted near human and we found that a pleasing female voice was to be the answer. Tastes may vary!!! We tried many voices but found that the Acapela voice sounded the best; we felt that the warbling type of voice that you get included with your operating system was not believable and, indeed, annoying. There are literally hundreds of voices to choose from, some are free and some are charged for.

The Acapela voice is a charged for voice, but it makes so much difference, your system becomes a believable "person"!! Worrying eh!?
Once you have decided on a voice and installed it on your operating system and set it to default, you must set the speed and pitch to your liking, this is done from within windows, control panel, speech app. Set it to too slow and the words don't sound right, set it too fast and every voice sounds like Donald Duck, default settings seem to work out very well.

Our Perl scripts call the default voice with its default settings; we can't alter them from within the perl server script.

Chapter 16

Making the system human

Getting a reply to your spoken commands.
If you are going to make your system more human, that is reply to just about any question or statement you make, the very first thing to do is set up a specific grammar file in order to give Rachel greater vocal scope in her recognition department.

With Rachel in command mode, as she is when switching lights etc, we use a "flat" grammar file, that is it has only to get her attention by using command word 'Rachel' and then she "listens" for the command, e.g. "Office light on".

If we are going to have Rachel interact with us, then we need to set up a tiered grammar structure. In a tiered grammar we still retain the Keyword for getting her attention "Rachel", and then we set up a sub key. The Sub Key is a word which in effect tells Rachel to look in a mini Grammar at a list only associated with the Sub Keyword, for example:

Rachel (Key Word) "Music"(Sub Keyword), at this stage "Rachel will look on the Sub Keyword list of music, where we would have other sub keywords and also commands.

Let me explain.
A sub Keyword here would be such as "Album" and under album we would have a list of album names, or we may have a command such as Play, Stop, Fast forward, Reverse, Volume Up, Volume Down.
We may even put in another layer under albums with keywords such as "Track"

Here is an example of a Grammar tree:

Chapter 16

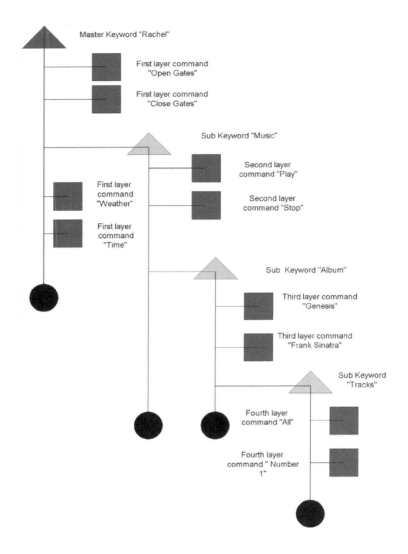

The diagram is just an example of how the grammar works, you can see from this tree like structure that the only word that will be recognised, initially is Rachel from there the only available command is "Hall Light on" or the Sub Keyword Music, so the spoken words "Rachel Music" would get you to the Music branch of the tree and from there you can give the command "Play" or "Rewind" or the Sub keyword "Album".

It should be clear that you cannot just say "Rachel Album" and expect to have the "album commands or sub keywords available. You have to follow the tree structure, so you would need to say "Rachel Music Album Genesis"

This enables the recognition engine to ignore all keywords that are not in the structure, this aids recognition as at any one point of the tree Rachel is only looking at the branch of the tree she is currently in.

Audio responses from the system control how Rachel speaks back to you.
It becomes apparent that if Rachel can be spoken to and she can "Reply" then we have the capability for interactive speech recognition. Now I won't pretend that this is easy, because it is not. You have to master the writing of XML Grammar files; the grammar we use to control Rachel is device specific and is not suited to conversational speech recognition.
However if you just want to experiment then you can add some commands to the config file to produce a basic interactive speech recognition grammar automatically.

Chapter 17

Using our system as a companion

It is not enough to have a dumb servant as it were; we are capable of much more.
When we designed Rachel, we wanted "her" to be able to respond to our commands by physically turning something on or off, but also to react as a human would; by giving audible feedback to the person giving the commands.
Thus if we say "Rachel Open Gates", it is real nice for Rachel to say, "I am opening the gates", or if we say "Rachel what is the time?" "She" can respond by saying "it is nearly twenty past 11, is there anything else I can do for you?"

This gets us into a whole new ballgame, we can talk to the computer and more importantly the computer can recognise what we say, act on it and reply in a very human way.
We do not even have to initiate the conversation, just write a perl script that Randomly chooses a time say between 9am and 6pm when she will just say" Hi this is Rachel is everything ok with you?" and if you say "Rachel, yes all ok" she will go back to sleep. If you don't reply she can ask again, and if no reply "she" can phone a friend or carry out some other predetermined action.

The possibilities are endless, let's go through a few and see what we get, starting with the Time Script. As we go through these examples I will detail what each new line does. By 'new', I mean if I have explained the purpose of a line in a previous example I shall ignore it, so if you don't see an explanation for a particular line, go back through the examples and you will find that line explained.

Time

This script gets local time from your PC and announces it in a humanised way:

So how does it work?

```
# This code is © R&DHarwood 2011. No part of this code may be copied, or used
unless  a copy of the book PC Voice Control System by R&D Harwood is owned by that
user, and this notice is preserved in the coding.
    0.   #!c:/perl/bin/perl.exe
    1.   use strict;
    2.   use Win32::OLE;
```

Chapter 17

```
3.   use Time::Human;
4.   my $voice;
5.   my $timeval = humanize(localtime());#speech starts here
6.   my $v=Win32::OLE->new('SAPI.SpVoice');
7.   $voice="Rachel(British)SAPI5";
8.   my @eee=$v->GetVoices();
9.   #print "voice1 $eee[0],voice2 $eee[1],voice3 $eee[2],voice4 $eee[3]";
10.  $v->{voice} = $v->GetVoices("name = $voice")->Item(0) if $voice;
11.  $v->Speak("The time is now ".$timeval);
12.  exit 0;
```

Start off as usual with the hash bang line **#!c:/perl/bin/perl.exe** which tells the system where the perl executable is located. (Note like all commented lines it does not require a semi colon at the end of the line).

We then instruct pearl to be harsh on us!! Yes, we want pearl to complain about everything it thinks isn't good programming practice, so we enter. This script requires the Time::Human module download and install from one of the perl repositories as shown in chapter 4 page 30

1. **use strict;**

2. **use Win32::OLE;**

3. **use Time::Human;** **This** simply takes a minute value and converts it to a vague time to or past the hour, for example 0619 would be converted to a text string "Nearly twenty past six".

It humanises the time in a manner that we would casually give the time to another person, when split second accuracy is not required.

This module was developed for text to speech scenarios like ours.

4. **my $voice;** is simply a declaration, remember we are using the "strict" pragma which demands that we declare variables before use(good Programming). In this variable we place the name of the SAPI voice we want our script to use, it can be the name of any installed sapi voice.

5. **my $timeval;** my $timeval is a declaration and it sets up the string to receive the humanised time value. The actual time is the PC's time "grabbed from the operating system in the variable "localtime"

6. **my $v=Win32::OLE->new('SAPI.SpVoice');** This is where we create a new OLE object wrapping the SAPI voice Engine.

7. **$voice = "Rachel(British)SAPI5";** Previously declared, we now assign a value to $voice, which is the name of the SAPI voice we want to use.

8. **my @eee=$v->GetVoices();** Declares and sets up an array of the sapi voices loaded on the PC system.

9. **# Print "voice1 $eee(0),voice2 $eee(1),voice3 $eee(2),voice4 $eee(3)";** This commented out line which if un- commented will print a list of voices installed on your PC.

10. **$v->(voice) = $v->GetVoices("name = $voice")->Item(0) if $voice;** This tells the win32::OLE to load the voice we have selected and then if the voice has loaded successfully, speak the time.

11. **$v->Speak("The Time is now. ".$timeval);** Tells Win32::OLE to speak the Time $textval.

12. **exit 0;** this line exits the program and returns a value of 0 for success.(we don't trap that returned value, but in certain circumstances you may wish to get that value to know the program ended successfully).

Weather

This is the Perl listing for getting Rachel to go onto the Internet, get and read the latest current weather for a particular area:
The first new line is **use Weather::underground;** download and install from one of the perl repositories as shown in chapter 4 page 30 this module gets meteorological information from the weather underground website (http://www.wunderground.com). It is a great source of meteorological information worldwide.

Following the next listing we describe in detail each part of the program.

Chapter 17

REMEMBER; we have included line numbers so that we can refer to each line, when describing the lines function.

When you are writing Perl DO NOT INCLUDE LINE NUMBERS, it won't work!

```
# This code is © R&DHarwood 2011. No part of this code may be copied, or used
unless   a copy of the book PC Voice Control System by R&D Harwood is owned by that
user, and this notice is preserved in the coding.

0.  #!c:/perl/bin/perl.exe
1.  use strict;
2.  use Weather::Underground;
3.  use Win32::OLE;
4.  my $v=Win32::OLE->new('SAPI.SpVoice');
5.  my $voice="Rachel22k";
6.  $v->{voice} = $v->GetVoices("name = $voice")->Item(0) if $voice;
7.  my $location="London, United Kingdom";
8.  my %wind={};
9.  $wind{"North"}="Northerly";
10. $wind{"NNE"}="North North Easterly";
11. $wind{"NE"}="North Easterly";
12. $wind{"ENE"}="East North easterly";
13. $wind{"East"}="Easterly";
14. $wind{"ESE"}="East South Eastly";
15. $wind{"SE"}="South Easterly";
16. $wind{"SSE"}="South South Easterly";
17. $wind{"South"}="Southerly";
18. $wind{"SSW"}="South South Westerly";
19. $wind{"SW"}="South Westerly";
20. $wind{"WSW"}="West South Westerly";
21. $wind{"West"}="Westerly";
22. $wind{"WNW"}="West North Westerly";
23. $wind{"NW"}="North Westerly";
24. $wind{"NNW"}="North North Westerly";
25. my $weather_obj = Weather::Underground->new(
            place => $location, cache_file=>'weather-now.txt',
                    cache_max_age=>120, debug => 0 )
            || die "Error, could not create new weather object:    $@\n";
26. my $weather = $weather_obj->get_weather() || die "Error, calling get_
    weather() failed: $@\n";
```

```perl
27. my %results={};
28. foreach (@$weather)
29. {
30. while ((my $key, my $value) = each %{$_})
31. {         $results{$key}=$value; }
32. }
33. announce("Hi its Rachel. Here is the latest weather for $location.");
34. announce("Weather timed at $results{updated}") unless $results{updated} eq
    "";
35. announce("Wind direction is $wind{$results{wind_direction}}") unless
    $wind{$results{wind_direction}} eq "";
36. announce("Wind speed is $results{wind_milesperhour} miles per hour") unless
    $results{wind_milesperhour} eq "";
37. announce("Humidity is $results{humidity} percent") unless
    $results{humidity} eq "";
38. announce("Current condition is $results{conditions}") unless
    $results{conditions} eq "";
39. announce("Sunrise today is $results{sunrise}") unless $results{sunrise} eq
    "";
40. announce("Sunset today is $results{sunset}") unless $results{sunset} eq "";
41. announce("Temperature is $results{temperature_celsius} degrees celcius")
    unless $results{temperature_celsius} eq "";
42. exit 0;

43. sub announce
44. {
45. my $text=shift;
46. print "$text\n";
47. $text=~s/ AM / A,M /;
48. $text=~s/Wind /Winde /;
49. $v->Speak(lc($text));
50. }
```

Start off as usual with the hash bang line **`#!c:/perl/bin/perl.exe`** which tells the system where the perl executable is located. (Note like all commented lines it does not require a semi colon at the end of the line).

We then instruct pearl to be harsh on us!! Yep we want pearl to complain about everything it thinks isn't good programming practice, so we enter. This script requires the Perl Weather::Underground module, download and install from one of the perl repositories as shown in chapter 4 page 30 .

Chapter 17

1. `use strict;`

2. `use Weather::Underground;`
Weather::Underground is a Perl module that, goes off into the void of the Internet and grabs the weather for your location from a site called www.wunderground.com. It returns all the readings we could want and makes them available to our script.

3. `use Win32::OLE;`
Explained earlier in Chapter 11.

4. `my $v=Win32::OLE->new('SAPI.SpVoice');`
Using Win32::OLE we create a Sapi.SpVoice object which will take our variables in text form and speak them later in the script

5. `my $voice="Rachel22k";`
This is a Declaration of variable $voice and assigning its value to the SAPI voice we want to use, in this case it's "Rachel22k"

6. `$v->{voice} = $v->GetVoices("name = $voice")->Item(0) if $voice;`
$v is assigned to pass voice type information to the SAPI engine

7. `my $location="London, United Kingdom";`
This Declaration sets the location that we want **www.wunderground.com** to get the weather for.

```
8.  my %wind={};
9.  $wind{"North"}="Northerly";
10.     $wind{"NNE"}="North North Easterly";
11.     $wind{"NE"}="North Easterly";
12.     $wind{"ENE"}="East North easterly";
13.     $wind{"East"}="Easterly";
14.     $wind{"ESE"}="East South Eastly";
15.     $wind{"SE"}="South Easterly";
16.     $wind{"SSE"}="South South Easterly";
17.     $wind{"South"}="Southerly";
18.     $wind{"SSW"}="South South Westerly";
19.     $wind{"SW"}="South Westerly";
20.     $wind{"WSW"}="West South Westerly";
21.     $wind{"West"}="Westerly";
22.     $wind{"WNW"}="West North Westerly";
23.     $wind{"NW"}="North Westerly";
24.     $wind{"NNW"}="North North Westerly";
```

Line 8 declares a hash and sets its values to lines 9- 24, and therefore changes the returned value that weather::underground retrieves for the wind direction from something like ESE to "East South East", which can be spoken by the SAPI Engine.

25. `my $weather_obj = Weather::Underground->new(place =>$location,cache_file=>'weather-now.txt',cache_max_age=>120, debug => 0 )|| die "Error, could not create new weather object:    $@\n";`

Line 25 Declares $weather_obj and assigns it to Weather::Underground, setting the location we desire and placing the data in a cache for later retrieval, if it fails it throws up the error message Error, could not create new weather object.

26. `my $weather = $weather_obj->get_weather() || die "Error, calling get_weather() failed: $@\n";`

Tells $weather to go get the weather and report an error message if it fails.

27. `my %results={};`

Sets up the %results hash.

28. `foreach (@$weather)`
29. `{`
30. `while ((my $key, my $value) = each %{$_})`
31. `{    $results{$key}=$value;   }`
32. `}`
33. `announce("Hi its Rachel. Here is the latest weather for $location.");`
34. `announce("Weather timed at $results{updated}") unless $results{updated} eq "";`
35. `announce("Wind direction is $wind{$results{wind_direction}}") unless $wind{$results{wind_direction}} eq "";`
36. `announce("Wind speed is $results{wind_milesperhour} miles per hour") unless $results{wind_milesperhour} eq "";`
37. `announce("Humidity is $results{humidity} percent") unless $results{humidity} eq "";`
38. `announce("Current condition is $results{conditions}") unless $results{conditions} eq "";`
39. `announce("Sunrise today is $results{sunrise}") unless $results{sunrise} eq "";`
40. `announce("Sunset today is $results{sunset}") unless $results{sunset} eq "";`

Chapter 17

```
41.         announce("Temperature is $results{temperature_celsius}
            degrees celcius") unless $results{temperature_celsius}
            eq "";
```
Lines 33-41 simply pass a text string and the contents of the %results hash to the announce sub routine, where the results are spoken. Unless the hash contains nothing, in which case it ignores the line.

```
42.         exit 0;
```
exit 0 literally exits the program and reports any error on exit.

```
43.         sub announce
```
This is the name of the Sub routine.

```
44.         {
```
Start the sub routine.

```
45.         my $text=shift;
```
Move the data into $text.

```
46.         print "$text\n";
```
Produce a view of the data in the command line console.

```
47.         $text=~s/ AM / A,M /;
```
To help the SAPI Engine pronounce words correctly, sometimes you have to alter the way they are written, so here we substitute AM for A comma M, to make the letters be spoken separately and not as "am" in spam.

```
48.         $text=~s/Wind /Winde /;
```
Same scenario as above but this time we want SAPI to say wind as in wind (gust of air) and not wind as in wind up the clock so we substitute with an "e" on the end to get the correct pronunciation.

```
49.         $v->Speak(lc($text));
```
Tell Sapi to go right ahead and speak the text, and when it's done go back to where it was called from. Note the lc(lower case) before($text)); that's there because SAPI treats upper and lower case differently. To get the correct inflection we need lower case in this situation.

Job done!!

Shares

This script will go get share prices and read them to you: ha ! and tell you how your fortune has diminished!! This script require the Finance::Quote module download and install from one of the perl repositories as shown in chapter 4 page 30.

This code is © R&DHarwood 2011. No part of this code may be copied, or used unless a copy of the book PC Voice Control System by R&D Harwood is owned by that user, and this notice is preserved in the coding.

```
0.  #!c:/perl/bin/perl.exe
1.  use strict;
2.  use Finance::Quote;
3.  use Win32::OLE;
4.  my $v=Win32::OLE->new('SAPI.SpVoice');
5.  my $voice="Rachel22k";
6.  $v->{voice} = $v->GetVoices("name = $voice")->Item(0) if $voice;
7.  my $quoter=Finance::Quote->new;
8.  my $exchange="nasdaq";
9.  my @symbols=('MSFT','GOOG','BHP');
10. my %info = $quoter->fetch($exchange,@symbols);
11. $v->Speak("your stocks and shares quotes.");
12. foreach my $stock (@symbols)
13. {
14. if (!$info{$stock,"success"})
15. {
16. print "Lookup of $stock failed - ".$info{$stock,"errormsg"}."\n";
17. $v->Speak("an error has occurred looking up price for $stock.");
18. next;
19. }
20. $v->Speak("the shares of ".$info{$stock,"name"}.", are currently trading at ".$info{$stock,"price"});
21. $v->Speak("the volume of shares is ".$info{$stock,"volume"});
22. }
23. exit 0;
```

Start off as usual with the hash bang line **#!c:/perl/bin/perl.exe** which tells the system where the perl executable is located. (Note; like all commented lines it does not require a semi colon at the end of the line.)

171

Chapter 17

We then enter:

1. **use strict;** Explained earlier in Chapter 11

2. **use Finance::Quote;** Finance quote is a perl module that goes off to the internet and retrieves stock market share prices and associated data. This data is accessed from the various stock exchange sites, but it should be noted that when stock exchanges give free data it is usually delayed by 15 minutes, so you will never get up to the minute values returned.

3. **use Win32::OLE;** This Perl module accesses Microsoft's OLE (Object Linking and Embedding, pronounced "olay") it lets one program use objects from another program.

4. **my $v=Win32::OLE->new('SAPI.SpVoice');** Using Win32::OLE we create a Sapi.SpVoice object which will take our variables in text form and speak them later In the script.

5. **my $voice="Rachel22k";** This is a Declaration of variable $voice and assigning it's value to the SAPI voice we want to use, in this case it's "Rachel22k".

6. **$v->{voice} = $v->GetVoices("name = $voice")->Item(0) if $voice;** $v is assigned to pass voice type information to the SAPI engine

7. **my $quoter=Finance::Quote->new;** This Declaration $quoter assigns a new object from the Finance::Quote module and exposes its values to us.

8. **my $exchange="nasdaq";** my $exchange (*no ! that's read Dollar Exchange*) sets the variable to the name of the stock exchange we wish to get information from.

9. **my @symbols=('MSFT','GOOG','BHP');** my @symbols Declares and sets an array containing the names of the stock market symbols we wish to access.

10. **my %info = $quoter->fetch($exchange,@symbols);** Here we Declare and fill a hash %info with the information we wish to access, using the $exchange and @ symbols variable's.

11. **$v->Speak("your stocks and shares quotes.");** This is where we start speaking, with an opening sentence calling the SAPI command speak.

12. **Through 18. foreach my $stock (@symbols)**

172

```
{
if (!$info{$stock,"success"})
{
print "Lookup of $stock failed - ".$info{$stock,"errormsg"}."\
    n";
$v->Speak("an error has occurred looking up price for
    $stock.");
```
next; This block cycles through the data for each share price requested and then if ok carries on. If no data found then it throws an error message "an error has occurred looking up price for $stock.");

19. } End curly braces.

20. `$v->Speak("the shares of ".$info{$stock,"name"}.", are currently trading at ".$info{$stock,"price"});` This is where we tell SAPI to speak the data we have requested on the stock name and stock price.

21. `$v->Speak("the volume of shares is ".$info{$stock,"volume"} });` And last but not least we get SAPI to speak the volume of the shares that we have requested.

22. } End curly braces.

23. `exit 0;` exit 0, literally exits the program and reports any error on exit.

There's your shares!

How to play music

There are a million and one programs out there that play music but we have gone for a freebie, and a very well known one at that; Winamp, (you can read about it later and we tell you where to download it).

Winamp suits our purpose because its design is flexible, it has a million skins available for it, and it just so happens that it has a control plug in, which is brilliant, called httpQ, written by Kosta Arvanitis.

httpQ brings all the capabilities of your Winamp player to our perl script program. httpQ is a tcp/ip server that listen s on a specified port for incoming requests.

Chapter 17

Using the custom httpQ protocol, users send requests to the server which in turn manipulates their Winamp player. Easy, eh?

Now if you're going to use this program, change line 5 to the ip address of your machine, or it won't work. Period. This script uses the IO::Socket::INET module download it from one of the Perl repositories and install as described in chapter 4 page 30.

```
# This code is © R&DHarwood 2011. No part of this code may be copied, or used
unless  a copy of the book PC Voice Control System by R&D Harwood is owned by that
user, and this notice is preserved in the coding.

    0.  #!c:/perl/bin/perl.exe
    1.  use strict;
    2.  use IO::Socket::INET;
    3.  use Win32::OLE;
    4.  my $socket=new IO::Socket::INET (
    5.  PeerHost => '192.168.0.192',
    6.  PeerPort => 4800, Proto => 'tcp')|| die "Unable to create socket: $!\n";
    7.  my $pw="rachel";
    8.  my $parms="";
    9.  my $cmd=@ARGV[0];
    10. if($cmd=~/(.*?)\&(.*)/)
    11. {
    12. $cmd=$1;
    13. $parms="&".$2;
    14. }
# build the string we are gonna send to the server

    15. my $urlstr = "/".$cmd."?p=$pw$parms";
    16. my $send="GET $urlstr HTTP/1.0\r\n\r\n";
    17. print $socket "$send";   # send it to the server
    18. my @recv=<$socket>;
    19. if($cmd eq "getplaylisttitle")
    20. {
    21. my $v=Win32::OLE->new('SAPI.SpVoice');
    22. my $voice="Rachel22k";
    23. $v->{voice} = $v->GetVoices("name = $voice")->Item(0) if $voice;
    24. $v->Speak($recv[4]);
    25. }
    26. $socket->close()||die"cant close";
    27. exit 0;
```

Chapter 18

Readily available hardware

As we stated earlier in the book there are many different boards available to carry out the PC to home automation interface. And indeed the average electrics enthusiast can build his/her own from scratch. The catch with building your own is that you need some pretty good programming skills to write an interface application or dll.
Velleman and PC-Control do some very good interface boards. We have limited ourselves to just those two manufacturers, not because the others aren't good enough, just they happened to be the easiest to interface to and use.

Interfacing to hardware K8055

Connecting the K8055 to your PC is a s simple as plugging in the USB cable (don't forget to load and register the k8055 DLL).

Connecting the K8056 is equally as easy, you need to connect all the outputs of the K8055 01-08 to the Inputs on the K8056, then you need to connect the ground lead from each, finally you need to provide around 12 volts ac to the K8056;a simple mains adapter will do the trick.

The K8055 does not require additional power for our project, it gets its power from the PC usb socket.

Interfacing to PC-control Master

Connecting the PC-control master (non isolated) is also just a case of plugging in a usb connection, and providing you have loaded the drivers that came with the master control unit , you are ready to go. With the isolated Master you require an additional 6 volt power supply in order to power the

Opto isolators.

The various Slave units simply connect via a two wire connection (although each slave does require a 6volt supply, either shared or its own), the connections on all the slave boards is parallel; A to A, B to B (plus there is a connection for screen if you are using shielded cables on long runs, remember you can place a slave as far as 1 kilometre away from the master!!).

Wired or wireless?

When we get down to the nitty gritty, we end up with a switch (relay) that is probably nowhere near where we want it to be, in regard to a light or power point.

You quickly realise that this project should have been done whilst the building was being constructed in order that you can get cables to where you want them.

Stand by here it comes!

IF YOU NEED TO RUN MAINS CABLES TO YOUR UNIT THEN GET A QUALIFIED ELECTRICIAN TO DO IT!!!

Nag! Nag! , Nag!

There are a number of wireless components out there, for switching power and light etc, just be aware that these are prone to interference (main culprit Cell phone, closely followed by the microwave oven), and it's really annoying if you get un-commanded operation of lights or power.

Chapter 19

How to control appliances

The control of appliances is specialised and we cannot go into it in detail here, so we will limit ourselves to explaining how to Control the Power socket.
To switch a socket on or off we need to make or break the live conductor (In the UK that's the wire having a Brown insulation sheath, it used to be Red).

To do this we take the live conductor from a fused power socket circuit and connect it to the live side of one of our relays and then the switched side of the relay is routed back to the Power socket.

Now a word of warning!

Think kettles for a minute.

Sure you can turn a kettle on and off, but when you turn it on, do you know for sure that it has water in it? If it doesn't you could be in for a burnt out element or worse! So we have to pick and choose what appliances we want to power up.
Then think about the implications of what happens when we turn on or off these appliances remotely!!

How to control gates

Gates are easy.
There's a bold statement, but its generally true in that most electric gates are fitted with a simple make or break connector to toggle the "Open Gates"-Close Gates " function.
This is normally provided to enable the fitting of a simple push button to open and close the gates.

All we need to do is to bridge that connection. It doesn't matter if there is a push button attached; we just bridge it in parallel.

So, by simply attaching the NO (Normally open) and Common contacts of the relay to the gates open switch we now have the gates under our control. If your gates don't have a Hold Open Function, then you will need to use the NC (Normally Closed) and common of another relay to provide the hold function by interrupting the power to the gate system, once you have sent the open command and with the gates are wide open. As with all automation, it is your health and safety responsibility to ensure that no person or object is in the path of the opening or closing gates!!

How to control doors

When we say controlling doors, what we have in mind is to be able to unlock a door using 'Rachel'. We are thinking of a scenario where a door is down a flight of stairs or some other location and you just want to be able to open it(hey make sure you know who wants to come in before you open it!).

So how do we do it?

We use an electric door lock, powered by 8-12 volts. These locks are suitable for surface and recessed mounting locks. We have them fitted to wooden doors but any type of material could be used.

We power this using a 9 volt transformer which can handle 1 amp.
The wiring circuit is easy; we take power from the transformer to the lock, taking any of the two leads via the contacts of a spare channel on the k8056. Add avery small modification to the perl server program to produce a four second pulse instead of our normal on and off on the relay.

Here is a very simple schematic:

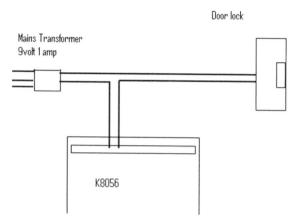

How to control lights

By the nature of the program it was born to control lights, but you need to think long and hard about mains voltages again. You are using potentially lethal voltages, so if you don't know what you are doing get the services of a qualified electrician.

So how did we make ours work?

Well we had an advantage in that we were wiring a building from scratch, so we were able to run all the light switch connections to one cupboard, where we had a number of k8056 boards.
Quite simply the live conductor to the light switch was taken via a relay on one of the K8056 boards. That K8056 board responded to the appropriate voice command to switch on or off the light.

The programs previously described already carry out this function so I do not need to reiterate, however a little schematic may be in order.

In the schematic I have shown one way switching and two way switching, if you are going to have two way switching then you must use the master slave relay board as they need to have a NO, NC and common connection to each relay.

Chapter 19

The schematic shows a low voltage lighting setup:

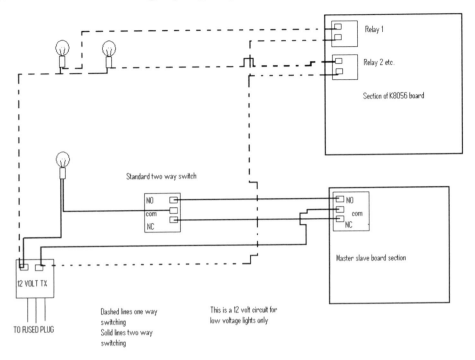

Chapter 20

Can Rachel tell me the temperature and humidity?

You just know the answer is yes, else we would not have posed the question, and in fact it's easy.

Temperature

If we are using the Velleman k8055 board we use the analogue inputs, a resistor (10K) and an NTC Thermistor (10k).

- Connect the NTC thermistor between A1 input and GND.
- Connect the 10k resistor between A1 and +5V of the K8055 board.
- Remove jumper from SK2. (You'll get the +5V from the other pin of the SK2 pin header.)

The values you get back are from 0 to 255. You will have to calibrate and set up a table to read the temperatures. You need to note the voltage over the NTC is very non-linear vs. temperature, so a little experimentation required.
Here is the perl script to get temperature values

```
# This code is © R&DHarwood 2011. No part of this code may be copied, or used
unless   a copy of the book PC Voice Control System by R&D Harwood is owned by that
user, and this notice is preserved in the coding.

use strict;
use Win32::API;
$|=1;
my $OpenDevice=Win32::API->new('K8055D','OpenDevice','I','I');
my $CloseDevice=Win32::API->new('K8055D','CloseDevice','','');
my $ReadAnalogChannel=Win32::API->new('K8055D','ReadAnalogChannel','I','I');
$OpenDevice->Call(0);

while(1==1)
{
my $AD1_value=$ReadAnalogChannel->Call(1);
```

Chapter 20

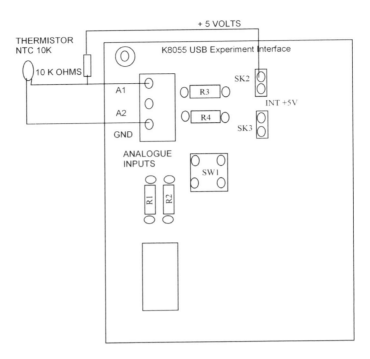

You will need to obtain values from your thermistor and then do a little maths to calibrate, but the principle is the same whatever temperature sensor you use.

print "rh=$rh%\r";
sleep 5;
}
$CloseDevice->Call();

exit 0;

This temperature script gets the raw analogue value, you must scale the value according to the data sheet of the particular thermistor or other analogue device you connect.

Humidity

We can sense humidity in a very similar fashion using Capacitive Humidity Sensor HIH4000-001 by Honeywell, which gives out values between 0 and 5 volts. Connect the sensor as shown with the +5 volts being picked up from the Sk2 Header. This device is quite linear with its outputs so calibration is pretty easy.

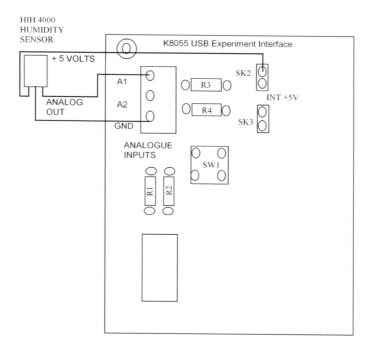

Perl script for humidity

This code is © R&DHarwood 2011. No part of this code may be copied, or used
unless a copy of the book PC Voice Control System by R&D Harwood is owned by that
user, and this notice is preserved in the coding.

Chapter 20

```perl
use strict;
use Win32::API;
$|=1;
my $OpenDevice=Win32::API->new('K8055D','OpenDevice','I','I');
my $CloseDevice=Win32::API->new('K8055D','CloseDevice','','');
my $ReadAnalogChannel=Win32::API->new('K8055D','ReadAnalogChannel','I','I');
$OpenDevice->Call(0);

while(1==1)
{
my $AD1_value=$ReadAnalogChannel->Call(1);
my $rh=sprintf("%3.2f",(((5/255)*$AD1_value)-0.958)/0.0309);
print "rh=$rh%\r";
sleep 5;
}
$CloseDevice->Call();

exit 0;
```

Exact readings will require additional work as the output from various forms of humidity sensor vary

Chapter 21

Can I have my cake and eat it?

Can I control the same system using a web interface as well as voice commands?
Yes of course you can. It is just a matter of writing a java or html page to send values to the perl server, and doing a little modification on the perl server script.
We hope to bring you this and other exciting scripts in a later volume.

Using mobile devices

There is no doubt that controlling the system from mobile devices, iphone, ipad etc is desirable. This subject could fill a small book on its own, but suffice to say we have the system running presently.

Chapter 22

Where can I get Perl?

Vanilla available free from Activestate, they do a free copy which comes with a good easy Package manager for getting additional modules. http://www.activestate.com/activeperl/downloads
Also you can download from Perl.org http://www.perl.org/get.html
And a great place for all things Perl is CPAN http://www.cpan.org/

Or perhaps you prefer Strawberry Perl for windows

Available from http://strawberryperl.com/
Both versions of Perl will work just fine, we just happen to like the package manager on activestate.

Where can I get VB6?

Well it's getting more difficult, because Microsoft in their wisdom decided to stop running with plain VB6 but insist that we all should use vb.net, or VB contained in visual studio, both of which are totally different animals, with accent on the animal bit.
But there is usually a copy or two of VB6 on ebay, either as a standalone copy or part of Visual Studio 6. The program will convert and run on VS2010

Where can I get TTS voices?

Acapela Group produces a very lifelike voice's check out their demo page at
http://www.acapela-group.com/text-to-speech-interactive-demo.html

If you want to buy these voices, you need to contact Steve Nutt.

Infovox Desktop distributor, Infovox Key distributor. Contact: Steve Nutt
- Website http://www.comproom.co.uk
- Email steve@comproom.co.uk
- Phone Number +44-1438-742286
- Fax +44-1438-759589 if you in the UK

or check out this link for other distributors http://www.acapela-group.com/distributors.php

It is interesting to note that Acapela do fine voices in twenty five different languages. We of course use the Rachel voice from their range. Hence our program is called Rachel!

Apart from Acapela there are other companies such as :
- 'Cepstral' http://www.cepstral.com/
- 'Ivona' do a good voice in "Emma" http://www.ivona.com/
- 'Next Up' sell tts voices from a number of companies http://www.nextup.com/?gclid=CP2g9ZvOlaYCFQkf4QodQxz2Zg

Where can I get hardware?

The Velleman K8055(VM110 pre built) and K8056 are available from a whole heap of retailers such as Maplin http://www.maplin.co.uk/module.aspx?moduleno=42857

ESR   Electronic Components http://www.esr.co.uk/velleman/

Rapid Electronics http://www.rapidonline.com/Educational-Products/Electronics/Velleman-Kits

And many more, A quick Google will get you a whole bunch of retailers. Shop around for the best price.

The PC-control interfaces can be purchased direct from http://www.pc-control.co.uk/. They have lots of information on their website and they deliver very promptly

Where can I get Winamp?

Winamp is a Media player, originally written mainly by a computer programmer named Justin Frankel back around 1997. He formed a company called Nullsoft which was subsequently acquired by AOL.

The success of Winamp was attributed to the fact that it was free/shareware, and had good visuals when playing back, could handle many types of music format (MP3, wav etc), and was skinable (that is, you could make it look as you wanted it). It had good playlist capability and best of all from our point of view some nice chap wrote a dll that could control Winamp from within different external programs.

Thus we had a music/ multimedia player under the control of a host program, in our case "Rachel".

You can download Winamp from http://www.winamp.com/media-player/download/en

At the time of writing all download sites mentioned were free of any detectable viruses...*But, just to make sure, have your antivirus turned on before you download!*

Index

#
#, 41, 42

2
230Vac, 146
240 volts, 19, 21

3
3 Phase, 19

A
Active Perl, 29
ActivePerl 27, 29
Activestate, 29, 187
Alexander Graham Bell, 15
alternating current, 22
Analogue to Digital, 146
API, 28, 106
Arrays, 42, 47
audio filters, 15
audio level, 34, 95
Audio level, 77, 94
audio oscillator device, 15
audio script, 40
automation, 16, 23

B
BASIC, 49
Bell Telephone Laboratories, 15
Binary, 84, 88
Bluetooth headset, 26
Burns, 21

Index

C
c:\rh2.xml, 92, 93
CGI scripting, 29
cmd, 31
command line, 41, 127
command prompt, 32, 92
Communications, 23, 25
Compiled, 41
Component Object Model, 1061
config file, 35, 37
Config File, 213
Configuration File, 61
Constants, 88

D
Darlington Transistor Array, 144
Date & Time, 58
DC shocks, 21
dch-serv3.pl, 92, 103
decimal to binary, 102
Declarations, 88
DHCP, 108
Dim statements, 55
dll, 32, 33
DLL, 32
Domestic appliances, 10, 23
DOS, 31, 32
Dynamic Link Libraries, 106, 107

E
earth, 19, 20
earth potential, 20, 21
else, 37, 45
ENIAC, 15
environmental path, 32
Error handling, 59
Event driven, 40, 89
experimenter's board, 32

Index

F
flow diagram, 65, 66
For loops, 46
foreach, 47, 167
Fourier Transform, 18, 199

G
Global, 42, 96
GNU Public License, 29
grammar, 35, 37
Grammar, 18, 35
grammar file, 37, 65
grammar xml file, 65, 140
GUI, 40, 49

H
hash-bang, 41
Hashes, 42, 102
hello world, 43, 50
Hello World, 31, 46
hello.pl, 31
home environment, 23
HTML, 35
humidity, 167, 169
hypothesis, 62, 94
Hypothesis, 39, 77

I
IBM, 15, 16
IDE, 41, 49
if statement, 44
Information Highway, 10, 13
Interpreted, 41
Interpreter, 41
ip address, 174
IP address, 108, 122
Isolated Master, 150

Index

K
k8055, 32, 33
K8055, 23, 27
K8055 Dll, 32, 107
K8055.dll, 27, 37
K8056, 23, 27
Kernel32, 89
Keyword, 37, 39

L
Launchdch-serv.bat, 91
LED, 146, 152
Lernout and Hauspie, 16
linguistic units, 16
Linux, 41
LocalHost, 108, 121

M
Mac, 26, 199
mains switch, 21
mains voltages, 21, 144
Maplin, 34
Markov chain, 199
Markov model, 18
Markov Model, 18, 199
mas. dll, 33
mas.dll, 27, 33
master controller, 149, 150
Master Slave units, 27
memory, 26, 98
microphone, 17, 26
Microsoft, 16, 27
Microsoft Common Dialogue Control 6(SP6), 74
Microsoft Internet Controls, 73
Microsoft SAPI 5.1, 27, 28
Microsoft Speech Object Library, 73
Microsoft Visual Studio 2010, 49
Microsoft Windows Common Controls 6.0 (SP6), 74
Microsoft Winsock Control 6.0, 74
Mixer, 34
Multimedia, 25, 189

N

Neurological effects, 21
NotePad ++, 31
Notepad++, 27, 37

O

OLE, 31, 106
OLE Automation, 73
OLE controller, 107
Open collector output, 144
Operators, 43, 56
Opto isolator, 150, 176

P

Parallels, 26
PC_CONTROL, 27
PC-Control Master, 149
PC-control master slave units, 33
perl, 29
Perl, 31, 41
Perl executables, 105
Perl modules, 43, 105
Perl Package Manager, 29, 105
perl server, 61, 62
Perl server, 65, 86
Perl Server, 35, 39
perl server script, 61, 63
personal computer, 16, 26
Phonemes, 17
phrase, 37, 39
PPM, 105, 106
pragma, 42, 106
preamp, 34
protective multiple earthing, 20

R

Rachel, 9, 25
Ray Kurzweil, 16
recognition Engine, 92, 93
Recognition Engine, 99
Recognition of voice, 39
relay, 62, 63

Index

relay kit, 27
Relay Slave, 149, 152
repositories, 29, 105
Repositories, 30
RS232, 146, 147
RS485, 149, 150
Rules, 35, 37

S
SAPI, 16, 28
SAPI 5.1, 27, 28
SAPI engine, 63, 66
sapi.dll, 28
SAPI5, 92, 105
SAVE, 70, 75
Scalars, 42
SDK, 28
Shares, 171
shebang, 41, 106
Shell, 90, 91
shock, 20, 21
socket, 25, 97
sound card, 17, 33
sound level, 74, 78
sound levels, 103
soundcard, 17, 26
speak.pl, 127, 128
Speech engine, 28
speech recognition, 13, 14
Speech Recognition engine, 28
status file, 66, 99
Status file, 140
status.dat, 108, 215
Status.dat, 215
Strawberry Perl, 187
String Functions, 56
stub file, 109
Stub File, 216
subroutine, 44, 45
subroutines, 45, 46
syntax, 41, 45

T

TCP connection, 200
TCP socket, 200
TCP/IP port number, 108
tcp/ip socket, 140, 141
TCP/IP socket, 105
text-to-speech, 157, 187
Text-To-Speech engine, 28
Thermistor, 181
training, 103
training the system, 23
transformer, 19, 20
TTL, 144
tts, 188
TTS, 157, 187

U

Unix, 29, 41
UNIX, 41, 199
URL, 199
usb, 26, 103
USB, 26, 32
use strict, 42, 106
use strict', 106

V

Variables, 42, 43
VB2010, 28
VB6, 27, 28
Velleman kit, 32
Ventricular fibrillation, 21
Vista, 27, 33
Visual Basic, 27, 28
Visual basic for Applications, 73
Visual Basic objects and procedures, 73
Visual Basic runtime objects and procedures, 73
Visual Studio, 28, 49
VM, 199
VM110, 188
vocabulary, 18, 37
Voice Recognition, 68
VU meter, 69, 74

W

Weather, 165
While loops, 46
Win32::API, 106, 107
Win32::GUI, 105
Win32::OLE, 32, 106
Win32::SAPI5, 105
Winamp, 173, 174
Windows 7, 27
Windows Operating System, 26, 27
winsock, 69, 90
Winsock, 75, 199

X

x86, 29
xml, 31, 35
XML, 35, 39
XP Professional, 27

Appendix A

Fourier Transform The Fourier transform is a mathematical operation that splits a signal(in our case wave file) into its constituent frequencies.

Markov chain or Markov Model , named after Andrev Markov, is a mathematical system that transits from one state to another (out of a countable number states) in a chainlike manner (in our case it produces a word from a collection of phonemes).

Mac Is the shortened name of an Apple Mac computer. Apple Inc. makes a series of computers that use a propriety operating system. In general terms it is not compatible with "IBM compatible PC's, although a number of crossover programs do exist, they would not suit our applications.

SAPI Speech Application Programming Interface is an Application Interface developed by Microsoft to use speech recognition and speech synthesis in Windows applications.

URL Uniform Resource Locator (URL) is a misused representation of an (URI) Uniform Resource Identifier. In computing, URL is commonly used to indicate where a resource may be found on the World Wide Web, such as www.google.com, but in fact that is a URI.

VM A virtual machine (VM) is a software implementation of a machine (i.e. a computer) that executes programs like a physical machine. A system virtual machine provides a complete system platform which supports the execution of a complete operating system such as windows.

Winsock is a programming interface that handles input and output requests for Internet applications in a Windows operating system. It's called Winsock because it is an adaptation of the Berkeley UNIX sockets interface for windows.
Sockets lays out a convention for connecting with and exchanging data between two programs within the same computer or across a network.
Sockets are used in our VB6 program using the winsock component.

Appendix A

TCP socket is defined as an endpoint for communication. A socket consists of the pair <IP Address, Port>. For our purposes, a port will be defined as an integer number between 0 and 65535. Although I should say that a lot of these ports are in use and have set numbers, such as http uses port 80. So in an ideal world we would use port numbers above 1024.

TCP connection consists of a pair of sockets. Sockets are distinguished by client and server sockets. A server listens on a port, waiting for incoming requests from clients.

Our Perl server program is a server. It sits and listens for commands from the VB6 program. Those commands are sent by the winsock **component of VB.

Appendix B Perl server listing

This code is © R&DHarwood 2011. No part of this code may be copied, or used unless a copy of the book PC Voice Control System by R&D Harwood is owned by that user, and this notice is preserved in the coding.

```perl
0.  #!c:/perl/bin/perl.exe
1.  use strict;
2.  use IO::Socket::INET;
3.  use Win32::API;
4.  use Win32::OLE;
5.  use Win32::Process;
6.  my $stub="";
7.  open(STUB,"<stub01.xml")||die" cannot open xml stub";
8.  while(<STUB>)
9.  {       $stub.=$_;      }
10. close STUB;
11. my @defines=();
12. my @phrases=();
13. my %cfg_val=();
14. my @config=();
15. my %lights=();
16. my %office=();
17. open(CFG,"<rachel-server.config")||die"cannot open rachel config";
18. while(<CFG>)
19. {
20. chomp;
21. if(/^#/) { next; } #ignore comment lines
22. push(@config,$_);
23. my @fields=split(/\|/);   # split the pipe delimited string into an array
24. if($fields[3] eq "Light")
25. { $lights{$fields[1].$fields[2]}=1; }
26. if($fields[4] eq "Office")
27. { $office{$fields[1].$fields[2]}=1; }
28. if($fields[0] eq "C") # if its a command feature, get tag & value pair
29. {
30.  $stub=~s/#$fields[1]#/$fields[2]/; # resolve it in the stub if present
31. next;
```

Appendix B Perl server listing

```perl
32. }
33. my $val="1$fields[1]$fields[2]"; # build a temp value
34. if($fields[0] eq "B") # if its a board control
35. {
36. my $state=1; # ON
37. my $state_word="on";
38. if($fields[3] eq "Switch")    { $state_word="open";
39. }
40. add_grammar_item($val,$state,$fields[5],$state_word);
41. $state=0; # OFF
42. $state_word="off";
43. if($fields[3] eq "Switch")    { $state_word="close";
44. }
45. add_grammar_item($val,$state,$fields[5],$state_word);
46. next;
47. }
48. if($fields[0] eq "S") # if its a script
49. {
50. add_grammar_item($fields[1],"",$fields[2],"");
51. next;
52. }
53. if($fields[0] eq "M") # if its a multi/macro
54. {
55. my $state=1; # ON
56. my $state_word="on";
57. add_grammar_item($fields[1],$state,$fields[2],$state_word);
58. $state=0; # OFF
59. $state_word="off";
60. add_grammar_item($fields[1],$state,$fields[2],$state_word);
61. next;
62. }
63. }
64. close CFG;
65. my $defines=join("",@defines);
66. $stub=~s/#DEFINES#\n/$defines/;
67. my $phrases=join("",@phrases);
68. $stub=~s/#PHRASES#\n/$phrases/;
69. open(GRAMMAR,">c:/rh2.xml")||die"cant open new grammar";
70. print GRAMMAR $stub;
71. close GRAMMAR;# thats the building of the grammar done...so lets load the last known status
72. my @status=();
```

Appendix B Perl server listing

```perl
73. open(STATUS,"<rachel-server.status")||die"cannot open rachel status";
74. while(<STATUS>)
75. {
76. chomp;
77. @status=split(/,/);
78. }
79. close STATUS;
80. my $OpenDevice=Win32::API->new('K8055D','OpenDevice','I','I');
81. my $CloseDevice=Win32::API->new('K8055D','CloseDevice','','');
82. my $WriteAllDigital=Win32::API->new('K8055D','WriteAllDigital','I','I');
83. my $port=5000;
84. my $socket = new IO::Socket::INET (
85. LocalHost => '192.168.0.03',LocalPort => $port,Proto => 'tcp',Listen => 5,Reuse => 1)||die "Unable to create socket on port $port : $!\n";
86. log_it("\n\nR&D Harwood 'RACHEL' Automation Server Initiated on port $port\n");
87. for (my $i=0;$i<4;$i++)
88. {
89. WriteAllDigital($i,$status[$i]);
90. log_it("initial startup, setting board $i to $status[$i]");
91. }# right, the boards have been 'set' to the values in the status file..# so, good to go.....listen for clients
92. my $HellFrozenOver = 0;
93. log_it("Listening for client....");
94. my $client_socket = $socket->accept();
95. my $peer_address = $client_socket->peerhost();
96. log_it("ah, there it is! $peer_address");# loop until the sun turns to a red giant and we are all sent to a fiery oblivion
97. while(!$HellFrozenOver)
98. {
99. my $data="";
100.        $client_socket->recv($data,1024);
101.        chomp($data);
102.        if ($data eq "7777") { $HellFrozenOver=1; last;
103.        }
104.        log_it("received: '$data'");
105.        my $response=&process($data);
106.        $client_socket->send($response."\n")||die"cant send back response";
107.        log_it("response sent - '$response'");
108.        log_it("waiting for next command");
109.        }
110.        $socket->close();
```

Appendix B Perl server listing

```perl
111.        exit 0;# main process finished, now some subroutines
112.        sub process
113.        {
114.        my $instruction=shift;
115.        my $time=localtime;
116.        if($instruction=~/^9/)
117.        {
118.        launch_ext($cfg_val{"$instruction"});
119.        return;
120.        }
121.        if($instruction=~/^8/)
122.        {   my $to_say="";
123.        if($instruction=~/^802/)
124.        {   $to_say="Office lights ";
125.        while (my ($key, $value) = each(%office))
126.        {
127.        action($key,substr($instruction,3,1));
128.        }
129.        }
130.        if($instruction=~/^801/)
131.        {   $to_say="All lights ";
132.        while (my ($key, $value) = each(%lights))
133.        {
134.        action($key,substr($instruction,3,1));
135.        }
136.        }
137.        if(substr($instruction,3,1) eq "0")
138.        { $to_say.="off";}
139.        Else
140.        { $to_say.="on";}
141.        my $Process_Object;
142.        Win32::Process::Create($Process_Object,"C:\\Perl\\bin\\perl.exe","perl speak.pl \"$to_say\"", 0,DETACHED_PROCESS, "." ) || die "cant speak";
143.        my $reply="";
144.        for (my $i=0; $i<4; $i++)
145.        {
146.        WriteAllDigital($i,$status[$i]);
147.        $reply.=$status[$i];
148.        if($i<3) {     $reply.=",";     }
149.        }
150.        write_status($reply);
```

Appendix B Perl server listing

```perl
151.        return $reply;
152.        }
153.        if (exists $cfg_val{$instruction})
154.        {
155.        my $to_say=$cfg_val{$instruction};
156.        $to_say=~s/\+//g;
157.        my $Process_Object;
158.        Win32::Process::Create( $Process_Object,"C:\\Perl\\bin\\perl.exe","perl speak.pl \"$to_say\"", 0,DETACHED_PROCESS, "." ) || die "cant speak";
159.        }
160.        my @parms=split(/,/,$instruction);
161.        my $reply="";
162.        if (uc($parms[0]) eq "STATUS")        {
163.        $reply=&status(\@parms);        } # pass array by reference
164.        if (uc($parms[0]) eq "INFO")   { $reply=&info(\@parms);
165.        } # pass array by reference
166.        if (uc($parms[0])=~/\d\d\d\d/)
167.        {       if(exists $cfg_val{"$parms[0]"})
168.        {
169.        my $type=substr($parms[0],0,1);
170.        my $board=substr($parms[0],1,1);
171.        my $channel=substr($parms[0],2,1);
172.        my $state=substr($parms[0],3,1);
173.        log_it("expands to: command type:$type, board num:$board, channel num:=$channel, bit state to:$state");
174.        my $temp=$status[$board];
175.        my $out = sprintf("%08d",unpack("B*", pack("N", $temp)));
176.        if(substr($out,7-$channel,1) eq $state)
177.        {
178.        print "already in that state\n";
179.        }
180.        Else
181.        {
182.        my $a=2**$channel;
183.        if($state eq "1")
184.        {       $temp+=$a;     } # detected ON request
185.        Else
186.        {       $temp-=$a;     } # detected OFF request
187.        $status[$board]=$temp;
188.        WriteAllDigital($board,$status[$board]);
189.        }
```

Appendix B Perl server listing

```perl
190.        $reply="";
191.        for (my $i=0; $i<4; $i++)
192.        {
193.        $reply.=$status[$i];
194.        if($i<3)
195.        {       $reply.=",";    }
196.        }
197.        write_status($reply);
198.        }
199.        Else
200.        {       print "Invalid command format, must be 4 digits - received '$parms[0]'\n";
201.        $reply="-1";
202.        }
203.        }
204.        return $reply;
205.        }
206.        sub log_it
207.        {
208.        my $time=localtime;
209.        my $msg=shift;
210.        $msg=~s/^(\n+)//;        # locate any extra blank lines at the start of a message
211.        my $blanks=$1; # and capture them to be used on the print. if we dont do this, we end up with blank lines between the timestamp and the message itself.
212.        print "$blanks$time - $msg\n";
213.        }
214.        sub status
215.        {
216.        my $parms=shift;  # receive array reference
217.        my $parm_count=@$parms;  # get number of array elements
218.        if ($parm_count > 1) { log_it("STATUS command received with unrequired parameters. ignoring them.");}
219.        my $reply="";
220.        for (my $i=0;$i<4;$i++)
221.        {
222.        $reply.=$status[$i];
223.        if($i<3){$reply.=",";}
224.        }
225.        return "$reply";
226.        }
```

```perl
227.        sub info
228.        {
229.        my $parms=shift;    # receive array reference
230.        my $parm_count=@$parms;   # get number of array elements
231.        if ($parm_count > 1) { log_it("INFO command received with unrequired parameters. ignoring them."); }
232.        my $return_info=join("#",@config);
233.        return $return_info;
234.        }
235.        sub OpenDevice
236.        {
237.        my $device=shift;
238.        my $rc=$OpenDevice->Call($device);
239.        }
240.        sub CloseDevice
241.        {
242.        my $rc=$CloseDevice->Call();
243.        }
244.        sub WriteAllDigital
245.        {
246.        my $device=shift;
247.        my $value=shift;
248.        OpenDevice($device);
249.        my $rc=$WriteAllDigital->Call($value);
250.        CloseDevice();
251.        }
252.        sub add_grammar_item
253.        {
254.        my $def=shift;
255.        my $state=shift;
256.        my $desc=shift;
257.        my $state_word=shift;
258.        $state_word=" +".$state_word unless $state_word eq "";
259.        if(exists $cfg_val{"$def$state"})
260.        {
261.        print "warning....value '$def$state' already exists. ignoring dupe\n";
262.        return;
263.        }
264.        $cfg_val{"$def$state"}=$desc.$state_word;
265.        push(@defines, qq`\t\t<ID NAME="DEF$def$state" VAL="$def$state" /> <!-- $desc$state_word -->\n`);
```

Appendix B Perl server listing

```perl
266.        $desc=~s/\&/\&amp;/g;
267.        push(@phrases, qq`\t\t\t<P VAL="DEF$def$state">$desc$state_word</P>\n`);
268.        }
269.        sub launch_ext
270.        {
271.        my $name=shift;
272.        my $parm="";
273.        if($name=~/(.*?)=(.*)/) { $name=$1; $parm=$2; }
274.        my $Process_Object;
275.        Win32::Process::Create( $Process_Object,"C:\\Perl\\bin\\perl.exe","perl $name.pl \"$parm\"", 0,DETACHED_PROCESS, "." ) || die "cant launch $name";
276.        }
277.        sub action
278.        {   my $item=shift;
279.        my $state=shift;
280.        my $board=substr($item,0,1);
281.        my $channel=substr($item,1,1);
282.        my $temp=$status[$board];
283.        my $out = sprintf("%08d",unpack("B*", pack("N", $temp)));
284.        if(substr($out,7-$channel,1) ne $state)
285.        {
286.        my $a=2**$channel;
287.        if($state eq "1")
288.        {       $temp+=$a;      } # detected ON request
289.        Else
290.        {       $temp-=$a;      } # detected OFF request
291.        $status[$board]=$temp;
292.        }
293.        }
294.        sub write_status
295.        {
296.        my $reply=shift;
297.        open(STATUS,">rachel-server.status") || die"cannot open rachel status for write";
298.        print STATUS $reply,"\n";
299.close STATUS;
300.}
```

Appendix C A sample of our grammar xml file

```xml
<GRAMMAR LANGID="409">
<DEFINE>
        <ID NAME="DEF1071" VAL="1071" /> <!-- Dummy0 +on -->
        <ID NAME="DEF1070" VAL="1070" /> <!-- Dummy0 +off -->
        <ID NAME="DEF1061" VAL="1061" /> <!-- Dummy1 +on -->
        <ID NAME="DEF1060" VAL="1060" /> <!-- Dummy1 +off -->
        <ID NAME="DEF1051" VAL="1051" /> <!-- Pool Table Area +on -->
        <ID NAME="DEF1050" VAL="1050" /> <!-- Pool Table Area +off -->
        <ID NAME="DEF1041" VAL="1041" /> <!-- Arcade Area +on -->
        <ID NAME="DEF1040" VAL="1040" /> <!-- Arcade Area +off -->
        <ID NAME="DEF1031" VAL="1031" /> <!-- Bottom of stairs +on -->
        <ID NAME="DEF1030" VAL="1030" /> <!-- Bottom of stairs +off -->
        <ID NAME="DEF1021" VAL="1021" /> <!-- Pool table +on -->
        <ID NAME="DEF1020" VAL="1020" /> <!-- Pool table +off -->
        <ID NAME="DEF1011" VAL="1011" /> <!-- Bar Area +on -->
        <ID NAME="DEF1010" VAL="1010" /> <!-- Bar Area +off -->
        <ID NAME="DEF1001" VAL="1001" /> <!-- Balcony +on -->
        <ID NAME="DEF1000" VAL="1000" /> <!-- Balcony +off -->
        <ID NAME="DEF1171" VAL="1171" /> <!-- Spot above bar +on -->
        <ID NAME="DEF1170" VAL="1170" /> <!-- Spot above bar +off -->
        <ID NAME="DEF1161" VAL="1161" /> <!-- Spot above tv +on -->
        <ID NAME="DEF1160" VAL="1160" /> <!-- Spot above tv +off -->
        <ID NAME="DEF1151" VAL="1151" /> <!-- Spot by seating +on -->
        <ID NAME="DEF1150" VAL="1150" /> <!-- Spot by seating +off -->
        <ID NAME="DEF1141" VAL="1141" /> <!-- Spot by door +on -->
        <ID NAME="DEF1140" VAL="1140" /> <!-- Spot by door +off -->
        <ID NAME="DEF1131" VAL="1131" /> <!-- Dummy4 +on -->
        <ID NAME="DEF1130" VAL="1130" /> <!-- Dummy4 +off -->
        <ID NAME="DEF1121" VAL="1121" /> <!-- Alcove left +on -->
        <ID NAME="DEF1120" VAL="1120" /> <!-- Alcove left +off -->
        <ID NAME="DEF1111" VAL="1111" /> <!-- Dartboard +on -->
        <ID NAME="DEF1110" VAL="1110" /> <!-- Dartboard +off -->
        <ID NAME="DEF1101" VAL="1101" /> <!-- Alcove right +on -->
        <ID NAME="DEF1100" VAL="1100" /> <!-- Alcove right +off -->
        <ID NAME="DEF1271" VAL="1271" /> <!-- Office six +on -->
        <ID NAME="DEF1270" VAL="1270" /> <!-- Office six +off -->
```

Appendix C A sample of our grammar xml file

```xml
<ID NAME="DEF1261" VAL="1261" /> <!-- Office five +on -->
<ID NAME="DEF1260" VAL="1260" /> <!-- Office five +off -->
<ID NAME="DEF1251" VAL="1251" /> <!-- Office four +on -->
<ID NAME="DEF1250" VAL="1250" /> <!-- Office four +off -->
<ID NAME="DEF1241" VAL="1241" /> <!-- Office three +on -->
<ID NAME="DEF1240" VAL="1240" /> <!-- Office three +off -->
<ID NAME="DEF1231" VAL="1231" /> <!-- Office two +on -->
<ID NAME="DEF1230" VAL="1230" /> <!-- Office two +off -->
<ID NAME="DEF1221" VAL="1221" /> <!-- Office one +on -->
<ID NAME="DEF1220" VAL="1220" /> <!-- Office one +off -->
<ID NAME="DEF1211" VAL="1211" /> <!-- Entrance +on -->
<ID NAME="DEF1210" VAL="1210" /> <!-- Entrance +off -->
<ID NAME="DEF1201" VAL="1201" /> <!-- Gates +open -->
<ID NAME="DEF1200" VAL="1200" /> <!-- Gates +close -->
<ID NAME="DEF9001" VAL="9001" /> <!-- weather -->
<ID NAME="DEF9002" VAL="9002" /> <!-- shares -->
<ID NAME="DEF9100" VAL="9100" /> <!-- music=play -->
<ID NAME="DEF9101" VAL="9101" /> <!-- music=pause -->
<ID NAME="DEF9102" VAL="9102" /> <!-- music=stop -->
<ID NAME="DEF9103" VAL="9103" /> <!-- music=prev -->
<ID NAME="DEF9104" VAL="9104" /> <!-- music=next -->
<ID NAME="DEF9105" VAL="9105" /> <!-- music=volumeup -->
<ID NAME="DEF9106" VAL="9106" /> <!-- music=volumedown -->
<ID NAME="DEF9107" VAL="9107" /> <!-- music=repeat&enable=1 -->
<ID NAME="DEF9108" VAL="9108" /> <!-- music=repeat&enable=0 -->
<ID NAME="DEF9109" VAL="9109" /> <!-- music=getplaylisttitle -->
<ID NAME="DEF8011" VAL="8011" /> <!-- All lights +on -->
<ID NAME="DEF8010" VAL="8010" /> <!-- All lights +off -->
<ID NAME="DEF8021" VAL="8021" /> <!-- Office Lights +on -->
<ID NAME="DEF8020" VAL="8020" /> <!-- Office Lights +off -->
<ID NAME="CmdType"    VAL="0001"/><!-- cant remove this else it doesn't compile so VAL first digit 0 =ignore-->
<ID NAME="Commands"   VAL="0002"/><!-- cant remove this else it doesn't compile VAL first digit 0 =ignore-->
</DEFINE>
<RULE ID="Commands" TOPLEVEL="ACTIVE">
<P>+Rachel</P>
<RULEREF REFID="CmdType" />
</RULE>
<RULE ID="CmdType" >
<L PROPID="CmdType">
<P VAL="DEF1071">Dummy0 +on</P>
```

Appendix C A sample of our grammar xml file

```xml
<P VAL="DEF1070">Dummy0 +off</P>
<P VAL="DEF1061">Dummy1 +on</P>
<P VAL="DEF1060">Dummy1 +off</P>
<P VAL="DEF1051">Pool Table Area +on</P>
<P VAL="DEF1050">Pool Table Area +off</P>
<P VAL="DEF1041">Arcade Area +on</P>
<P VAL="DEF1040">Arcade Area +off</P>
<P VAL="DEF1031">Bottom of stairs +on</P>
<P VAL="DEF1030">Bottom of stairs +off</P>
<P VAL="DEF1021">Pool table +on</P>
<P VAL="DEF1020">Pool table +off</P>
<P VAL="DEF1011">Bar Area +on</P>
<P VAL="DEF1010">Bar Area +off</P>
<P VAL="DEF1001">Balcony +on</P>
<P VAL="DEF1000">Balcony +off</P>
<P VAL="DEF1171">Spot above bar +on</P>
<P VAL="DEF1170">Spot above bar +off</P>
<P VAL="DEF1161">Spot above tv +on</P>
<P VAL="DEF1160">Spot above tv +off</P>
<P VAL="DEF1151">Spot by seating +on</P>
<P VAL="DEF1150">Spot by seating +off</P>
<P VAL="DEF1141">Spot by door +on</P>
<P VAL="DEF1140">Spot by door +off</P>
<P VAL="DEF1131">Dummy4 +on</P>
<P VAL="DEF1130">Dummy4 +off</P>
<P VAL="DEF1121">Alcove left +on</P>
<P VAL="DEF1120">Alcove left +off</P>
<P VAL="DEF1111">Dartboard +on</P>
<P VAL="DEF1110">Dartboard +off</P>
<P VAL="DEF1101">Alcove right +on</P>
<P VAL="DEF1100">Alcove right +off</P>
<P VAL="DEF1271">Office six +on</P>
<P VAL="DEF1270">Office six +off</P>
<P VAL="DEF1261">Office five +on</P>
<P VAL="DEF1260">Office five +off</P>
<P VAL="DEF1251">Office four +on</P>
<P VAL="DEF1250">Office four +off</P>
<P VAL="DEF1241">Office three +on</P>
            <P VAL="DEF1240">Office three +off</P>
<P VAL="DEF1231">Office two +on</P>
<P VAL="DEF1230">Office two +off</P>
<P VAL="DEF1221">Office one +on</P>
```

Appendix C A sample of our grammar xml file

```xml
        <P VAL="DEF1220">Office one +off</P>
        <P VAL="DEF1211">Entrance +on</P>
        <P VAL="DEF1210">Entrance +off</P>
        <P VAL="DEF1201">Gates +open</P>
        <P VAL="DEF1200">Gates +close</P>
        <P VAL="DEF9001">weather</P>
        <P VAL="DEF9002">shares</P>
        <P VAL="DEF9100">music=play</P>
        <P VAL="DEF9101">music=pause</P>
        <P VAL="DEF9102">music=stop</P>
        <P VAL="DEF9103">music=prev</P>
        <P VAL="DEF9104">music=next</P>
        <P VAL="DEF9105">music=volumeup</P>
        <P VAL="DEF9106">music=volumedown</P>
        <P VAL="DEF9107">music=repeat&amp;enable=1</P>
        <P VAL="DEF9108">music=repeat&amp;enable=0</P>
        <P VAL="DEF9109">music=getplaylisttitle</P>
        <P VAL="DEF8011">All lights +on</P>
        <P VAL="DEF8010">All lights +off</P>
        <P VAL="DEF8021">Office Lights +on</P>
        <P VAL="DEF8020">Office Lights +off</P>
        </L>
        </RULE>
</GRAMMAR>
```

Appendix D   Config File listing

```
# This code is © R&DHarwood 2011. No part of this code may be copied, or used
unless  a copy of the book PC Voice Control System by R&D Harwood is owned by that
user, and this notice is preserved in the coding.
# C - Command feature
# B - Board control  No Audio confirmation
# S - Script
# M - Multi/Macro
C|COMMAND_WORD|+Rachel
B|0|7|Light|Games Room|Dummy0
B|0|6|Light|Games Room|Dummy1
B|0|5|Light|Games Room|Pool Table Area
B|0|4|Light|Games Room|Arcade Area
B|0|3|Light|Games Room|Bottom of stairs
B|0|2|Light|Games Room|Pool table
B|0|1|Light|Bar|Bar Area
B|0|0|Light|Balcony|Balcony
B|1|7|Light|Bar|Spot above bar
B|1|6|Light|Bar|Spot above tv
B|1|5|Light|Bar|Spot by seating
B|1|4|Light|Bar|Spot by door
B|1|3|Light|Bar|Dummy4
B|1|2|Light|Games Room|Alcove left
B|1|1|Light|Games Room|Dartboard
B|1|0|Light|Games Room|Alcove right
B|2|7|Light|Office|Office six
B|2|6|Light|Office|Office five
B|2|5|Light|Office|Office four
B|2|4|Light|Office|Office three
B|2|3|Light|Office|Office two
B|2|2|Light|Office|Office one
B|2|1|Light|Games Room|Entrance
B|2|0|Switch|Garden|Gates
S|9001|weather
S|9002|shares
S|9100|music=play
S|9101|music=pause
```

Appendix D Config File listing

```
S|9102|music=stop
S|9103|music=prev
S|9104|music=next
S|9105|music=volumeup
S|9106|music=volumedown
S|9107|music=repeat&enable=1
S|9108|music=repeat&enable=0
S|9109|music=getplaylisttitle
M|801|All lights
M|802|Office Lights
```

Appendix E Status.dat

The file status.dat is not big, its list is the next line..

0,0,12

and that's it, just copy and save it in your perl directory as status.dat

Appendix F Stub File

This is the listing for stub.xml it must be verbatim and saved in the perl directory as stub.xml

```
<GRAMMAR LANGID="409">
        <DEFINE>
#DEFINES#
                <ID NAME="CmdType"    VAL="0001"/><!-- cant remove this else it
doesn't compile so VAL first digit 0 =ignore-->
        <ID NAME="Commands"   VAL="0002"/><!-- cant remove this else it doesn't
compile VAL first digit 0 =ignore-->
        </DEFINE>
        <RULE ID="Commands" TOPLEVEL="ACTIVE">
                <P>#COMMAND_WORD#</P>
                <RULEREF REFID="CmdType" />
        </RULE>
        <RULE ID="CmdType" >
                <L PROPID="CmdType">
#PHRASES#
                </L>
        </RULE>
</GRAMMAR>
```